MATTERING PRESS

Mattering Press is a scholar-led Open Access publisher that operates on a not-for-profit basis as a UK registered charity. It is committed to developing new publishing models that can widen the constituency of academic knowledge and provide authors with significant levels of support and feedback. All books are available to download for free or to purchase as hard copies. More at matteringpress.org.

The Press' work has been supported by: Centre for Invention and Social Process (Goldsmiths, University of London), European Association for the Study of Science and Technology, Hybrid Publishing Lab, infostreams, Institute for Social Futures (Lancaster University), OpenAIRE, Open Humanities Press, and Tetragon Publishing.

We are indebted to the ScholarLed community of Open Access, scholar-led publishers for their companionship and extend a special thanks to the Directory of Open Access Books and Project MUSE for cataloguing our titles.

MAKING THIS BOOK

Books contain multitudes. Mattering Press is keen to render more visible the otherwise invisible processes and people that make our books. Our gratitude goes to our readers, for books are nothing without them and our supporters for helping us to keep our commons open. We thank the editors and contributors, and the reviewers Jennifer Gabrys, Harald Rohracher, Daniel Worden – and a fourth who opted to remain unnamed. Further, we thank Steven Lovatt for the copy-editing; Alice Ferns for the manuscript formatting; Tetragon Publishing for the typesetting and design; Neil Ford for the cover illustration; Julien McHardy for design; Will Roscoe for our website and for maintaining our books online; Anna Dowrick for caring for the book's promotion and for its community of readers and contributors; and Julien McHardy and Natalie Gill who acted as the production editors for this book.

ENERGY WORLDS

In Experiment

EDITED BY

JAMES MAGUIRE

LAURA WATTS

BRIT ROSS WINTHEREIK

Mattering Press

First edition published by Mattering Press, Manchester.

Cover art © Neil Ford, 2021
Cover design © Julien McHardy, 2021
Text designed and typeset by Tetragon, London

Freely available online at matteringpress.org/books/energy-worlds

ISBN: 978-1-912729-08-1 (pbk)
ISBN: 978-1-912729-09-8 (ebk)
DOI: https://doi.org/10.28938/9781912729098

Mattering Press has made every effort to contact copyright holders and will be glad to rectify, in future editions, any errors or omissions brought to our notice.

This book was made possible with funding by The Danish Independent Research Council.

CONTENTS

LIST OF FIGURES

ARTWORK

CONTRIBUTORS

SIMONE ABRAM is Professor in Anthropology at Durham University, and co-director of the Durham Energy Institute. She was a co-investigator at the National Centre for Energy Systems Integration from 2016–2021, and from 2020 is a co-investigator at the Research Centre for Inclusive Decarbonisation led by Tanja Winther at Oslo University. Relevant recent publications include *Electrifying Anthropology: Exploring Electrical Practices and Infrastructures* (Bloomsbury, with Brit Ross Winthereik and Thomas Yarrow), Our Lives with Electric Things (in *Cultural Anthropology*, with Jamie Cross, Lea Schick and Mike Anusas) and *Ethnographies of Power* (Berghahn, with Tristan Loloum and Nathalie Ortar).

MÓNICA AMADOR-JIMÉNEZ is a Research Associate at the School of Geographical Sciences at the University of Bristol in the interdisciplinary research project 'BioResilience: Biodiversity Resilience and Ecosystem Services in Post-Conflict Socio-Ecological Systems in Colombia'. Ph.D. Candidate in Anthropology at the University of Oslo, with research experience on armed and environmental conflicts in Colombia. Master's in gender and cultural studies from the University of Chile, and professional experience at UNHCR and UNDP related to advocacy and research about Colombian refugees, asylum seekers and migrants in South America.

ANDREA BALLESTERO is Associate Professor of Anthropology at Rice University. She is also the founder and director of the Ethnography Studio (https://ethnographystudio.org/). Her research examines spaces where the law, economics and techno-science are so fused that they appear identical. Her areas of interest include the politics of knowledge production; economic, legal and political anthropology; water politics; subterranean spaces and liberalism. She is the author of *A Future History of Water* (Duke, 2019). With Brit Ross Winthereik she co-edited *Experimenting with Ethnography: A Companion to Analysis* (Duke 2021). She is now conducting research on how aquifers, property, and science yield new spatial imaginaries of the underground in Costa Rica.

GEOFFREY C. BOWKER is Donald Bren Chair at the School of Information and Computer Sciences, University of California at Irvine, where he directs the Evoke Laboratory, which explores new forms of knowledge expression. Recent positions include Professor of and Senior Scholar in Cyberscholarship at the University of Pittsburgh iSchool, and Executive Director, Center for Science, Technology and Society, Santa Clara. Together with Leigh Star he wrote *Sorting Things Out: Classification and its Consequences*; his most recent books are *Memory Practices in the Sciences* and (with Stefan Timmermans, Adele Clarke and Ellen Balka) the edited collection: *Boundary Objects and Beyond: Working with Leigh Star*. He is currently completing a trilogy of papers on ants, fungi and global consciousness and writing a book on time and computing.

DOMINIC BOYER teaches at Rice University where he also served as Founding Director of the Center for Energy and Environmental Research in the Human Sciences (2013–2019). He is currently pursuing anthropological research with flood victims in Houston, Texas, and on electric futures across the world. His most recent book is *Energopolitics* (Duke, 2019), which is part of a collaborative duograph, "Wind and Power in the Anthropocene," with Cymene Howe, which studies the politics of wind power development in Southern Mexico. With Howe, he also helped make a documentary film about Iceland's first major glacier (Okjökull) lost to climate change, *Not Ok: A Little Movie about a Small Glacier at the End of the World* (2018). In August 2019, together with Icelandic collaborators they installed a memorial to Okjökull's passing, an event that attracted media attention from around the world.

JAMIE CROSS is Professor of Social and Economic Anthropology at the University of Edinburgh. He is the author of *Dream Zones: Anticipating Capitalism and Development in India* (Pluto Press, 2014) His writing on the social and material politics of off grid solar energy in places of chronic poverty has been published in *South Atlantic Quarterly*, *Limn*, the *Journal of the Royal Anthropological Association*, *Ethnos* and *The Guardian*. His collaborations with designers, visual artists, and filmmakers include: *Solar What?!* (an award winning, fully repairable, open source solar powered lighting and charging device); the *Off Grid Solar Scorecard* (a public platform to track sustainable design in the solar industry); and *The Solar Fix* (a short film about solar things in need of repair).

ENDRE DÁNYI is Visiting Professor at the Department of Sociology at the J.W. Goethe University in Frankfurt am Main, Germany, and University Fellow at the Charles Darwin

University in Darwin, Australia. Endre's long-term interest is in places and material practices associated with democratic politics. His PhD research was a material-semiotic analysis of the Hungarian Parliament, whereas his Habilitation investigates various blind spots of parliamentary politics, including the European refugee crisis, the 'War on Drugs' and Indigenous initiatives in Northern Australia.

NEIL FORD makes comics, prints and stuff for the web. He lives & works in Orkney, 59th parallel North. @neil_ford.

REBECCA FORD is a postgraduate research student and teaching assistant with the University of the Highlands and Islands (UHI), based at the Institute for Northern Studies in Orkney, and a director on the board of the Orkney Renewable Energy Forum (OREF). Her research is informed by the work of Mikhail Bakhtin and her experience of community, growing up and living in Orkney. In her PhD project, 'Words and Waves: A Dialogical Approach to Discourse, Community, and Marine Renewable Energy in Orkney', she develops ecological dialogism – an approach to language and meaning-making as an embodied process enacted within a physical and cultural environment – to understand the role of narrative and the process of community in the development of Marine Renewable Energy in Orkney. In 2014 Rebecca worked as a Research Assistant for the Alien Energy project at the IT University of Copenhagen, carrying out fieldwork at the European Marine Energy Centre (EMEC) in Orkney.

STEFAN HELMREICH is Professor of Anthropology at MIT. He is the author of *Alien Ocean: Anthropological Voyages in Microbial Seas* (University of California Press, 2009) and of *Sounding the Limits of Life: Essays in the Anthropology of Biology and Beyond* (Princeton University Press, 2016). His essays have appeared in *Critical Inquiry*, *Representations*, *American Anthropologist*, and *The Wire*.

CYMENE HOWE is Professor and Director of Graduate Studies in the Department of Anthropology at Rice University. She is the author of *Intimate Activism* (Duke 2013) and *Ecologics: Wind and Power in the Anthropocene* (Duke, 2019), an ethnographic study of renewable energy infrastructures and their cultural and environmental impacts in Oaxaca, Mexico. She is co-editor of *The Anthropocene Unseen: A Lexicon* (Punctum, 2020), and *The Johns Hopkins Guide to Critical and Cultural Theory*. Her current research, Melt/Rise, focuses on climate adaptation and the relationship between cryospheric loss in the Arctic and sea level rise in lower latitude coastal cities. With Dominic Boyer, she produced the documentary film *Not Ok: A Little Movie about a Small Glacier at the End of the World* (2018) and in August 2019, initiated the installation of the world's first memorial to a glacier felled by climate change. The Okjökull memorial event (https://en.wikipedia.org/wiki/Okjökull) in Iceland was meant as a global call to action and in memory of a world rapidly melting away.

ROB JONES is a writer and Letterer of comics. His work has been published in such companies as Image Comics, Heavy Metal, Humanoids, DC Thomson, Scout Comics, Behemoth Comics, BHP Comics and many more. He is based in Yorkshire in the UK and enjoys various types of sandwiches. He can often be found shouting madness into the void on twitter at @RobJonesWrites or at the pigeons whilst in his dressing gown at the local park.

ANN-SOFIE KALL is a researcher and teacher at the School of Education and Communication, Jönköping University. She has an interdisciplinary background within social science and humanities, and a PhD in technology and social change from Linköping University. Her work combines inspirations from STS, sociology, political science and history. Her research is mainly focused on relations between politics, technology and nature, with a special interest in energy- and environmental politics. In the ongoing research project, 'The World Needs a New Narrative' she, together with her colleagues, explores the use of narratives and stories in relation to environmental issues and the transformation of society.

HANNAH KNOX is Associate Professor of Anthropology at UCL and Director of the UCL Centre for Digital Anthropology. Her work focuses on the study of technical projects as sites of social and cultural change, and she has conducted ethnographic research in the UK and Latin America. Recent books include *Roads: An Anthropology of Infrastructure and Expertise* (co-authored with Penny Harvey) and *Ethnography for a Data Saturated World* (co-edited with Dawn Nafus) and *Thinking like a Climate: Governing a City in Times of Environmental Change* (Duke, 2020).

JAMES MAGUIRE is Assistant Professor at the IT University of Copenhagen. His work focuses on the manifold interfaces between, and within, environmental and digital concerns. His current book project is an ethnographic exploration of the temporal and political consequences of energy extraction in Iceland. His ongoing research is oriented towards sustainable digitalization; an enquiry into how digitalization has become an object of attention for sustainable thinking. This involves projects that explore the paradoxical relationship between the deleterious environmental effects of digital processes and their promissory imaginaries of climate mitigation, as well as those that speculate about, and activate, alternative ways of creating more ethically inflected digital futures.

NOORTJE MARRES is Professor in Science, Technology and Society and Director of the Centre for Interdisciplinary Methodologies, University of Warwick. Her work investigates issues at the intersection of innovation, everyday environments and public life: the role of mundane objects in environmental engagement, intelligent technology testing in society, and changing relations between social life and social science in a computational age. She also contributes to methodology development, in the area of issue mapping. Noortje studied sociology and philosophy of science and technology at the University of Amsterdam, and is currently a Visiting Professor in the Centre for Science and Technology Studies (CWTS) at the University of Leiden. She published *Material Participation* (Palgrave, 2012) and *Digital Sociology* (Wiley, 2017), and with Michael Guggenheim and Alex Wilkie edited *Inventing the Social* (Mattering Press, 2018). More information at www.noortjemarres.net.

DAMIAN O'DOHERTY is Professor of Management and Organization at the Alliance Manchester Business School where he is director of the Manchester Ethnography Network and co-founding director of BEAM – the nuclear and social science research network, at the Dalton Institute of the University of Manchester. Damian is an organization theorist who works ethnographically in organizations, broadly defined, through which he seeks to reanimate latent politics at play in organizations. He has published widely in the academic press and currently serves as associate editor for the journal Organization and was former editor in chief for Culture and Organization (2008–2011).

LEA SCHICK is a researcher investigating innovation for sustainable futures at the intersection of art, design and infrastructure studies. Lea is studying smart grids and how 'consumers' are imagined to become more active and involved in their own energy consumption through home displays and economic incentives. Engaging with disciplines such as art, design and architecture she is proposing more social and creative ways of engaging citizens in energy and sustainability. She teaches sustainable IT and management at the IT University of Copenhagen, and works as an innovation specialist at the Alexandra Institute, where she is working with sustainable transitions in SMEs.

MICHAELA SPENCER is a Research Fellow with the College of Indigenous Futures, Arts and Society at Charles Darwin University, Australia. Her current research involves working with

Indigenous knowledge authorities, and differing traditions of knowledge and governance. This involves collaborative research for policy development, and engaging with government, service providers, university staff and Indigenous people in remote communities. So far, this work has focused on issues such as disaster resilience, emergency management, governance and leadership, remote engagement and coordination, volunteering, health and wellbeing.

LAURA WATTS is a writer, poet, ethnographer of futures, and Senior Lecturer in Energy & Society within Geosciences, University of Edinburgh. As a science and technology studies (STS) scholar, her research is concerned with the effect of 'edge' landscapes on how the future is imagined and made, along with an exploration of different writing methods. For the past decade she has been working with people and places around energy futures in the Orkney islands, Scotland. Her latest book *Energy at the End of the World: An Orkney Islands Saga* (MIT Press) was shortlisted for the Saltire Research Book of the Year, and she won the International Cultural Innovation Prize 2017, as part of the Reconstrained Design Group, for a community-built energy storage device designed from spare parts. For more on her work see www.sand14.com.

BRIT ROSS WINTHEREIK is full Professor at the IT University of Copenhagen in the Technologies in Practice group and head of the Center for Digital Welfare. She has published on public sector digitalisation, information infrastructures, and ethnography for anthropology and STS audiences. She is co-author of *Monitoring Movements in Development Aid: Recursive Infrastructures and Partnerships* (MIT Press, 2013) with Casper Bruun Jensen, and co-editor of *Electrifying Anthropology: Exploring Electrical Practices and Infrastructures* (Bloomsbury, 2019) with Simone Abram and Thomas Yarrow, and of *Experimenting with Ethnography: A Companion to Analysis* (Duke, 2021) with Andrea Ballestero. She is part of the Anthropology of Technology network which edits *Handbook for the Anthropology of Technology* (Palgrave Handbook Series, 2022). She is a frequent participant in public debates on issues related to public digitalization, and appointed member of the Digital Advisory Council for the Academy of the Technical Sciences in Denmark. She was PI of the Alien Energy project (2013–2016) and of the Data as Relation project (2017–2020).

ACKNOWLEDGEMENTS

We are grateful to all those involved with the 'Alien Energy' project that ran from 2013–2017. The chapters in this collection came into being during a workshop at the IT University of Copenhagen in October 2017. We are thankful for the participants' willingness to engage with our ideas and collaboration formats. The project's advisory board consisted of Per Ebert (consultant), Peter Karnøe (Professor at Aalborg University), Jan Krogh (Danish Wave Energy Center), John Law (Professor Emeritus at The Open University) and Marianne Elisabeth Lien (Professor at the University of Oslo). We are grateful for the time, care and insight they put into working with us. Also involved in the project were PhD student Louise Torntoft Jensen, Research Assistant Line Marie Thorsen, Research Assistant Simon Carstensen and Rina de Place Bjørn from the IT University's communications department. Thanks for your time and effort in supporting us.

Everyone who took part in the Energy Walk, as well as our field site partners in Denmark, Orkney and Iceland, are acknowledged for their welcoming cooperation and friendship. Our families have been providing us with tremendous support through the full process of the project. Finally, we wish to acknowledge The Danish Independent Research Council for providing funding for the research.

FOREWORD:
SETTING THE SCENE

THE PREPARATION FOR THIS BOOK BEGAN IN COPENHAGEN. LET'S RETURN THERE FOR a moment and imagine a group of scholars in a light and airy university space. Drawing pads and pencils are stuffed onto bookshelves. Electronic components, circuit boards and pamphlets lie around in a casual half-order that communicates creativity. It is a room with a designerly feel to it; furniture can be switched around whenever there is a need to alter our group working constellations.

The people in the room are scholars at different stages of their careers. They have travelled from afar to share their research on energy, the environment and climate change. As organisers, we had asked them to do something unusual; namely, to present their research with an eye to how their individually crafted ideas and theoretically curated interests could be recomposed into new, multi-authored texts. The new texts would take as their starting point themes that this newly formed collective considered absent, or underarticulated, within anthropology and social studies of energy. All workshop participants have collaborated creatively and intensely to produce this volume.

The workshop was the conclusion to a research project called *Marine Renewable Energy as Alien: Social Studies of an Emerging Technology*, that ran from January 2013 to December 2016 and was funded by the Independent Research Fund Denmark (ID 0602-02115B). As part of the project, the editors conducted ethnographic studies of water-based renewable energy – wave, tide and geothermal – at sites around the North Atlantic. Using science and technology studies (STS) and anthropology as our approaches to this research, and in our dissemination of it, we have sought to open up renewable energy to scrutiny as we followed the work and concerns of engineers, geologists and community innovators in Iceland, Orkney and Denmark (Maguire 2019, 2017; Maguire and Winthereik 2017; Watts and Winthereik 2018; Winthereik, Maguire, and Watts 2019; Watts 2019; Winthereik 2019).

Both in our own research and in the workshop from which this book emerged, we couple various notions of energy together with various notions of experiment. In working with these themes through the re-composition of the texts, this coupling became the orientation point for our collaborative post-workshop discussions and writings. A key point of this book is that energy and experiment must go together in the work of figuring new energy futures, including how such futures connect with what has gone before.

The coupling of energy and experiment emphasises processes of bringing forth, of *making*. It also reflects the authors' interest in science and technology in the making of alternative energy futures. And it explores the possibilities of playful scholarly engagement, while remaining grounded within our academic traditions. 'Experimental' – as we deploy it in this volume – denotes an approach whereby our empirical and analytical methods both form and are formed by our object of study. We share with anthropologists of creative improvisation the belief that experimentation and creative practice are shared across worlds – ours and the worlds of those whom we study (Ingold and Hallam 2007).

There are two related meanings of the word 'experimental' that we want to put across here. In common parlance, to be experimental means to be playful, but we want to be more systemic in the specific ways in which playfulness can meet scientific traditions of experimentation. The effort of tacking back and forth between different experimental sites (from laboratories, to landscapes, to university workshops) is one characteristic of the approach to the experimental making of energy worlds adopted in this volume. We explore the experimental as a process of making energy worlds through comparison (Deville, Guggenheim and Hrdličková 2016). The concept – and title of the book – *Energy Worlds*, alludes to how the making of renewable energy futures seems to happen in places that experiment with energy through relational contestation. The contribution of this book is an effort to render energy worlds through an openness towards collective writing as an infrastructure for social inquiry in an area that deserves both scholarly and political attention.

REFERENCES

Deville, J., M. Guggenheim, and Z. Hrdličková, eds, *Practising Comparison* (Manchester: Mattering Press, 2016).

Ingold, T., and E. Hallam, eds, *Creativity and Cultural Improvisation* (Oxford: Berg, 2007).

Maguire, J., 'Icelandic Geopower: Accelerating and Infrastructuring Energy Landscapes'. In 'Technologies in Practice' (PhD Thesis, IT-University of Copenhagen, 2017).

— 'The Temporal Politics of Anthropogenic Earthquakes: Acceleration, Anticipation, and Energy Extraction in Iceland', *Time and Society* (27 November 2019) <https://doi.org/10.1177/0961463X19872319>.

Watts, L., *Energy at the End of the World: An Orkney Islands Saga* (Cambridge, MA: MIT Press, 2019).

Watts, L. and B. R. Winthereik, 'Ocean Energy at the Edge', in G. Wright, S. Kerr, and K. Johnson, eds, *Ocean Energy: Governance Challenges for Wave and Tidal Stream Technologies* (London: Routledge, 2018), pp. 229–246.

Winthereik, B. R., 'Is ANTs Empiricism Ethnographic?', in A. Blok, I. Farias and C. Roberts, eds, *The Routledge Companion to Actor-Network Theory* (Abingdon and New York: Routledge, 2019).

Winthereik, B. R., L. Watts, and J. Maguire, 'The Energy Walk: Infrastructuring the Imagination', in J. Versi and D. Ribes, eds, *digitalSTS: A Field Guide for Science & Technology Studies* (Princeton and Oxford: Princeton University Press, 2019), pp. 349–363.

I

INTRODUCTION

James Maguire, Laura Watts and Brit Ross Winthereik

ENERGY IS EVERYWHERE AND NOWHERE; AN ETHEREAL MATERIAL-SEMIOTIC (Anusas and Ingold 2015), both alien and familiar. Quotidian, at times sublime, energy's significance in the formation of industrialised societies over the last century is hard to dispute (Gupta 2015). As anti-extractivist ('Keep it in the Ground') campaigns and student movements gain momentum the world over, the Paris Climate Agreement continues to lag behind more recent climate science estimates of a sustainable world. At the same time, alternate political configurations are emerging on the ground, as individual cities and communities band together to act upon their collective concerns, bypassing national regulations and implementing their own renewable energy schemes (Blok and Tschötschel 2016).

Yet while renewable technologies for harnessing wind, wave, solar and geothermal energy advance and prices fall, political representatives seem to be at an impasse (Szeman et al. 2016). The Intergovernmental Panel for Climate Change (IPCC) has suggested clear lines of action on reducing emissions, but the political capacity for taking such action still seems lacking. This suggests that questions of energy are far more than questions of technology or policy: they are about values and ethics, culture, power and institutions (Boyer and Szeman 2014).

As citizens and scholars, we are situated in the middle of a disconcerting conundrum: the legacy of *homo sapiens* as a species has become so extensive as to have altered the planet's geophysical processes, yet our traditions and conventions of political deliberation are still unfit for managing our planetary alterations (Chakrabarty 2017). While geo-engineering solutions are gaining pace, our institutional and political capacities are slowing by comparison. This pushes us to enquire about, and participate in, the making of renewable energy worlds.

This introduction will outline the histories and lineages that shape this volume. It is intended as a situated accounting of energy research in anthropology, science and technology studies (STS) and related fields, without making any claim to be exhaustive. We consider the landscape of energy literature on which we draw to be more topological than topographical (Strathern 1991), in the sense that we bring in bodies of work that allow for journeys across both energy research in anthropology/STS and also across different expressive formats.[1] The scholars assembled in this book share a commitment to exploring ways of writing and communicating that seek to bring out otherwise invisible aspects of energy and energy research. They have co-authored six sections, each guided by a specific technique and ethos. The emergent work is indebted to several fields of research and situated within a collective of individuals who share a particular concern: how to animate the making of energy futures by experimental means.

The experimental form of the book – constructed through text, graphics and interactive materials – can be mobilised in various ways: as a tool for thinking energy otherwise; as a means of energy engagement and critique; or simply as a playful artefact that stimulates temporary energy imaginaries and collectives. The following pages provide a series of routes through energy literature that link the book to contemporary energy research in the fields of STS and anthropology, and beyond.

ENERGY RESEARCH IN ANTHROPOLOGY AND STS

Energy infrastructures and the public understanding of energy have long been a concern in studies of science, technology and innovation, while more recently anthropology has once again reengaged with the topic (Bolton, Lagendijk and Silvast 2018; Schick and Winthereik 2013; Shove and Walker 2014; Silvast 2017a, 2017b; Silvast, Hänninen and Hyysalo 2013; Sovacool 2014). Dominic Boyer's book *Energopolitics* (2019) outlines the discipline's 'sporadic engagement' with energy. From the thermodynamically inclined universal theory of energy proposed by Leslie White in the 1950s, to the more indigenously sensitive accounts of the 1970s, to the recent flurry of publications on the anthropology of energy (Behrends and Reyna 2011; Cross 2014; Howe and Boyer 2015a; McNeish and Logan 2012; Nader 2010; Strauss, Rupp and Love 2013; Winther and Wilhite 2015; Wylie 2018), this sporadic impulse to study energy is spurred by vulnerable or transitional moments in the dominant political regimes of energy.

Boyer's move to determine anthropology's engagement with energy is twofold. Firstly, he pivots to Leslie White, who makes a deterministic, neo-evolutionary link between a society's energy use and its degree of cultural development. Instead of a return to determinism, however, Boyer signals a trend in anthropological studies of energy towards integrating a more materialist politics. This materialism is of a different kind to White's. It attends not only to the culture and politics of energy – the approach advocated by Strauss et al. in *Cultures of Energy* (2013) – but also suggests a method for energy studies that pays attention to the emergence of political configurations through the materiality of energy forms. Interestingly, these analytical moves follow the contours of the socio-material analytics of Timothy Mitchell's *Carbon Democracy* (2009), a book which charts the relationship between the twentieth century's carbon abundance and the development of particular infrastructures of democratic politics. In his ground-breaking work, Mitchell (2009) skilfully compares coal, oil and wood in terms of the political formations they afford, connecting modes of energising society with modes of governing it. It is this analytical mode that Boyer outlines in *Energopolitics* – the elicitation of energy's politics through its materiality by, 'searching out signals of the energo-material transferences and transformations incorporated in all socio-political phenomena' (Boyer and Szeman 2014: 325).

In the volume *Infrastructures and Social Complexity*, the editors point towards the ways in which energetic materials are implicated in broader issues of sociality and politics (Harvey, Jensen and Morita 2017). They emphasise the entangled relations between these materials, technologies, and forms of politics that give rise to new modes of sociality. As they neatly put it; 'the differential powers of water or sun, (for example), inflect not only the temporalities and spatialities of energy, but also the subjects we may be able to become' (Harvey, Jensen, and Morita 2017: 158). This observation suggests that to engage in energy scholarship today means being attentive to issues of force, flow and matter as they intersect with, or generate, energy politics. This is one challenge that this volume will take up: an attention to how the materialities of energy afford political possibilities, as well as to how politics make energy practices, technologies, and infrastructures manifest.

STS, in particular, has a longstanding tradition of taking up questions concerned with the materialisation of energy through technology. The edited volume *Electrifying Anthropology* tackles the question of how electricity can become a matter of concern in the social sciences and humanities and brings together theoretical resources and cases studies from STS and anthropology. One of the main issues here is: What languages must we adopt and develop to be able to account for the presence of electric forces and flows, and their impact on social life? (Abram, Winthereik, and Yarrow 2019)

More classic examples range from studies of electric cars, to engagements with public transport systems, to electricity meters in the Côte d'Ivoire, to the meanings associated with electric lightbulbs.[2] Recent sustainable energy research has taken a more consumer-based approach to these questions by focusing on the devices and technologies that make energy consumption more visible, and thereby more amenable to behavioural intervention (Burgess and Nye 2008; Hargreaves et al. 2010). More resonant with the experimental mode of this book is Jennifer Gabrys' *A Cosmopolitics of Energy* (2014). Gabrys asks if it is possible to open up our thinking on energy materialities by pushing beyond approaches that emphasise engagement with energy consuming devices and monitoring technologies – approaches that render materiality in physicalist terms. Gabrys argues for an Isabelle Stengers-inspired line of enquiry that creates cosmopolitical connections across energy collectives. As we will see in Chapter 3, 'Propositional politics', the politics of water distribution, geothermal engineering practices and the design of interactive spaces are brought into a form of dialogue that resonates with this sense of cosmopolitics. This method holds open the possibility of hesitation, i.e. of working through speculative registers that attend to the generative and yet-to-be-known capacities of energy practices (Gabrys 2014, 2015). Here, materialisations of energy are connected to the politics of participation as new collectivities and practices are given form through experimental, creative interventions.

The current energy impasse has also been addressed by sociologist John Urry (2014), who frames it as 'the problem of energy'. For Urry, the problem is not just the lagging pace of carbon transitions, but that energy has been, and continues to be, a problem for social theory itself. As with the scholars above, Urry has suggested that the ways in which societies are energised are crucial for how they work, noting how various forms of energy work to structure the social, temporal and spatial organisation of society, and life itself. Social theory, Urry argues, has paid limited attention to energy systems, including the practices and habits they generate (Urry 2014: 4–5). Take, for example, Zygmunt Bauman's (2000) 'liquid modernity', one of the twentieth century's most influential concepts. Social theory has been almost blind to how such liquid relations have been dependent upon 'black gold' (oil; Urry 2014: 6).

Another sociologist and STS scholar, Andrew Barry (2015: 110–111), takes up this challenge, querying why the *concept* of energy is not more prominent in both studies of energy and studies of materiality. His argument, briefly put, is not to suggest that energy should provide a new basis for rethinking materiality, but to open up the question of how scientists understand energy's political import as it is converted between different forms (Barry 2015: 111). Adopting Isabelle Stengers' thermodynamic view of the relationality of matter, Barry suggests that matter is neither inert nor lively, but is always already a part of ongoing energetic relations. As such, his turn to Stengers is a way for him to link *energy as concept* with *energy as political practice*.[3] While Urry takes social science to task for its energy myopia – both in terms of its historical and conceptual blind spots – Barry pushes us to think of energy concepts as part of a more materialist politics.

Another scholarly site of promise is the emergent 'energy humanities' (Boyer and Szeman 2014; Pinkus 2016; Szeman et al. 2016; Szeman and Boyer 2017; Szeman, Wenzel, and Yaeger 2017; Wilson, Carlson, and Szeman 2017). This field of research brings together scholars from within the humanities, social sciences and beyond, in an effort to grapple with the complexities of energy. Energy humanities asks what, if anything, social and cultural scholars can do to render a world in which renewable energy transitions become more likely. In the process, it blurs the boundaries between academic disciplines, and between academic and applied research. Of particular interest to this book is the way in which these scholars articulate questions of energy-aesthetics as part of the understanding, analysis and performance of our current climate change impasse. Energy humanities ponders why it is that literary and artistic representations of energy are almost absent from our contemporary cultural imaginaries. They seek to bring together artful renderings with scholarly concerns. Prompted by Amitav Ghosh's (1986) essay 'Petro-fiction' – which asks why encounters with oil have proven so imaginatively sterile as to have failed to develop a body of fictional work – energy humanities has begun to examine the borders between energy, culture and representation. Ghosh's response to his own provocation is to suggest that scholars must confront oil's 'slipperiness,' such that it is brought back within the frame of our critical atten-tion. Energy's 'unseen' status within great aesthetic works as the great 'not-said' is a silence that can no longer be ignored (Szeman et al. 2017: 18).

This point is, of course, not new to those familiar with STS literature, which has for many years articulated methods for making visible the quotidian mundanities of infrastructures (Bowker and Star 1999; Edwards et al. 2009; Star and Ruhleder 1996). With a few excep-tions (Jensen 2018), scholars within STS are less concerned with aesthetics and literary reproduction, and more inclined towards analyses that bring attention to the scientific and techno-political work needed to stabilise energy systems, socially, institutionally and politically. However, recent work from public engagements with science and technology takes up the question of aesthetics in the making of energy collectives. For example, using speculative design, STS scholars and design researchers Alex Wilkie and Mike Michael deploy playfully designed artefacts to explore energy technologies in everyday settings, with a focus on these artefacts' aesthetic effects, along with their potential political role in opening up (or closing down) new energy communities and futures (Wilkie and Michael 2018; Boucher et al. 2018). It seems clear that some variants of STS and energy humanities share a performative and experimental sensibility.

The current energy impasse is part of the material, political and conceptual 'problem of energy', and requires scholarly attention attuned to addressing energy's ambiguities as a material thing, political concern and analytical concept. Breaking the impasse requires experimental modes of research and writing, which we point towards in the book's title: *Energy Worlds in Experiment*. However, as already mentioned, a comprehensive or holistic review of energy literature is beyond the remit of this book, which means that some difficult, yet necessary, choices have been made that exclude various literature sets. Take, for example, concerns around the type of renewable energy being developed in Iceland – one of the ethnographic field sites evoked in this book. The extraction of geothermal energy is producing anthropogenic earthquakes not dissimilar to those that result from fracking (Maguire 2019). While the constellation of issues generated through fracking are pressing – questions around the politics of health and risk, indigenous rights, scientific knowledge production and expertise, as well as more classical political economy questions as to the role of 'big energy companies' – this literature is not front and centre in this volume. The same can be said of the body of work that engages with energy rights and ethics, energy violence and justice, as well as the transition literature and literature on energy sustainability. The topics dealt with in this book indicate the particular interests and expertise of the authors, and a broad perspective on what might count as energy research in STS-anthropology has been our main focus. Consequently, we engage primarily with research that addresses everyday encounters with energy, energy infrastructures as complicated political-material accomplishments, materialities of energy, expertise around energy in formation, and existing and emerging cultures of energy.

EXPERIMENTS IN FORM: INTRODUCTION TO THE CHAPTERS

Although this book is experimental – in both content and form – it is also a critical engagement with the politics of energy futures. But such critical engagement involves, according to Donna Haraway, taking the inseparability of fiction and fact seriously. To Haraway (2016), crafting academic and scientific knowledge requires speculative fabulation as a form of serious play – science fact is also, always, science fabulation. Haraway (2015) inspires us to bring out more variation, more entry points, more registers and modalities to studies of energy. In this volume we submit that fabulation and speculative fiction are the *sine qua non* of systematic explorations of energy worlds (Tsing 2010).

We feel there is an ethical as well as a methodological imperative to engage researchers in rendering different energy worlds (Blok, Nakazora and Winthereik 2016; Holbraad and Pedersen 2017). Given the longstanding attention to writing as experiment in the fields of STS and anthropology, this volume builds on an established tradition. Its forms are part of analytical writing techniques employed by scholars in other contexts (Tsing and Pollman 2005; Latour 1986; Mol 2002; Raffles 2011). All of the contributing authors attend to the varying potentials of writing as embedding modes of performativity. The reader will encounter familiar and less familiar forms of academic text, narrative, proposition, theses, artwork, card game and interview. As a whole, the authors' experiments in form resonate with John Law's argument for a 'method assemblage', i.e. being conscious of how we relate to – and circumvent – the apparatus of analysis in order to be responsible for its world-making effects, multiple accountabilities and politics (Law 2004).

Each chapter of the book employs a different experimental format. Following this introduction, Chapter Two by Ann-Sofie Kall, Rebecca Ford and Lea Schick, uses a narrative format akin to extended ethnographic vignettes. This format invites the reader into a series of embodied experiences of energy infrastructures, each seen through the eyes of a first-person narrator. As an effect of this narration style, challenges in low carbon energy are localised in practice at field sites in Orkney, Sweden and Denmark. Attention to these sites and their forms of energy helps the authors weave together their experiences.

The authors use material artefacts that they encountered during ethnographic fieldwork as prompts to weave their stories of energy infrastructures together. The artefacts invite us to reflect on how each energy world is told as a narrative, which travels to, and so connects with, other places and worlds. The weaving of stories through found objects is the experimental form that this chapter takes, and its analysis presents a subtle argument about the presence of the global in the local, and the mobility of energy and energy stories. The focus on stories channels the argument that, as energy researchers, we have a responsibility to make our narrative techniques and analytical devices – in this case, the prompts – visible, in order to be accountable to the changes that we would like to see happening.

In Chapter Three, Endre Dányi and Michaela Spencer, James Maguire, Hannah Knox and Andrea Ballestero likewise show what has gone into the making of the chapter and its argument. The key concept and title of the chapter is 'propositions', and its central point is that we need to think of new, more imaginative ways to reinvigorate our politics as the planet enters a threshold phase of impending environmental and social calamity. The chapter argues for a mode of politics (propositional politics) that takes into account the voices of various silent Others. The experimental nature of this chapter lies in its performative form, being constructed as a series of case stories that build upon one another. The authors generate reflections on each other's texts through an intra-textual mediation that charts the analytical move from experiment to proposition – with varying gradations in between. In essence, the text is an exercise in practising what it preaches by performing its argument through the construction of the text. The chapter points out that, rather than search for ontological clarity and unambiguous responsibility, scholars should explore ties, even less obvious ones, between practices and logics that seem incommensurate. As a result, we can learn to acknowledge the inherently contradictory and ambivalent consequences of political projects pertaining to energy.

Chapter Four, written by Brit Ross Winthereik, Stefan Helmreich, Damian O'Doherty, Mónica Amador-Jimenéz and Noortje Marres, borrows its form from philosophical texts that seek to carve out a space for political intervention though brevity. We speak here of the thesis as a form of writing (not doctoral, think Luther or Feyerabend). The chapter revolves around the making of an energy polity: a communal gathering with political consequence. Following a short introduction which gestures at connections between energy infrastructures and forms of governance, each author provides an ethnographic narrative which addresses the question: What are the conditions for the formation of an energy polity in this particular environment? The types of energy that the contributions to this chapter describe are not all renewable. Weaving the various stories together, the authors express a deep concern with, and interest in, how different forms of energy enact community. Whether these communities finally become energy polities is not definitively answered. The chapter considers the energy polity as a political form *in becoming*; and while it is not a form that has quite managed to coalesce, it is nevertheless shown to be political. The experimental form of thesis-writing embraces this ambiguity by making the extent to which the thesis has political agency an open question. This openness is performed by a list of five theses, which can be cut out of the book and used in whatever way the reader sees fit.

Chapter Five, by Laura Watts, Cymene Howe and Geoffrey C. Bowker, takes the form of a graphic novel. This chapter explores the boundaries of the academic text as a particular epistemic form that can be moulded, stretched and transformed into a graphic style. The making of the chapter was informed by the notion of 'interruptions'. What is being interrupted here is both the readers' expectation of what an academic text contribution should look like, and perhaps also the authors' hope that readers will know how to engage with a graphic contribution. The chapter asks its readers to learn to read the form in order to get to know Unda, the main character of the novel. As a figure in the text, Unda invites the attuned reader to embark upon a journey with her across continents. It is on this journey that different forms of energy are encountered and experienced through Unda as an energetic companion.

In Chapter Six, Simone Abram and Jamie Cross invite readers to engage in energy matters through the form of a card game – ElectroTrumps – which offers alternative histories of energy. The details on the cards are taken from classic energy experiments, as well as particular tests and trials in the development of electricity. Each card elicits questions about what might otherwise be seen as the esoteric production of scientific knowledge and technology. The game is modelled on 'Top Trump' cards from the 1960s, and by inviting players to invent their own cards, it materialises the changing relationship between science, technology and the public. As the authors write in the framing text: 'ElectroTrumps sets up new opportunities for collectively engaging in juxtaposition, comparison and reflection'. This, to us, indicates a twofold ambition: to reclaim the specificities of electrical experiments and their histories, and to develop novel means of constructing opportunities for imagining our future lives with electricity. The chapter links to a website from which the ElectroTrump cards and templates for new cards can be downloaded. In doing so, the authors invite readers to engage students, colleagues, friends and family in figuring out electricity pasts, presents and futures. Through what may be seen as the quotidian politics of a card game, the chapter shows how actors from the Global South have made significant contributions to scientific and technical renderings around electricity.

Chapter Seven is a conversation between anthropologist Dominic Boyer and one of the editors, James Maguire, focusing on Boyer's scholarly and activist interests in energy and the Anthropocene. The interview addresses questions around the analytical forms being generated within studies of energy in anthropology-cum-STS collectives, as well as the politics of intervention in a world fraught with institutional and political gridlock. The conversation challenges the very notion of what constitutes both intervention and the legitimacy of particular modes of anthropological inquiry. It ends on a note of cautious optimism, by interrogating the emerging imaginaries around hyper-electric, hyper-local energy worlds.

The final chapter is a short conclusion in which the editorial collective reflect upon where the volume lands us in terms of further exploring energy worlds and the experimental.

NOTES

1 Many of those gathered in this collection self-ascribe to these fields, yet we recognise that the category anthropology-STS can never be a holistic identity. The hyphenated category is itself one which is as broad as it is contested. Rather, the collection has gathered ANT-inspired STS scholars with an ethnographic proclivity alongside anthropologists with a bent towards experimentation.

2 See Callon 1986; Latour 1996; Akrich 1992; Bijker 1995. These cases come courtesy of Wilkie and Michael 2018: 129.

3 Barry is interested in Isabelle Stengers' concept 'cosmopolitics', particularly the relationship it describes between scientific practices and politics. Barry claims that while Stengers' work is not overtly political, her intent is to reorient the hierarchy of the sciences so as to undo the dominance of physics in understandings of matter. Influenced by Stengers, Barry uses energy conversion and measurement as a way to address ongoing questions of materiality and energy in STS – in particular, how physical scientists transform reality in order to render energy into a calculable and comparable form.

REFERENCES

Abram, S., R. Winthereik and T. Yarrow, *Electrifying Anthropology: Exploring Electrical Practices and Infrastructures* (New York and London: Bloomsbury, 2019).

Akrich, M., 'The De-scription of Technical Objects', in W. Bijker and J. Law, eds, *Shaping Technology/Building Society: Studies in Sociotechnical Change* (Cambridge, MA: MIT Press, 1992), pp. 205–24.

Anusas, M., and T. Ingold, 'The Charge against Electricity', *Cultural Anthropology*, 30.4 (2015): 540–554.

Ashmore, M., *The Reflexive Thesis: Wrighting Sociology of Scientific Knowledge* (Chicago: University of Chicago Press, 1989).

Barry, A., 'Thermodynamics, Matter, Politics', *Distinktion: Scandinavian Journal of Social Theory*, 16.1 (2015): 110–125.

Bauman, Z., *Liquid Modernity* (Cambridge: Polity, 2000).

Behrends, A., and S. P. Reyna, 'The Crazy Curse and Crude Domination: Towards an Anthropology of Oil', in A. Behrends, S. P. Reyna, and G. Schlee, eds, *Crude Domination: An Anthropology of Oil* (New York: Beghahn Books, 2011), pp. 3–29.

Benjamin, W., *The Arcades Project* (Harvard: Harvard University Press, 1999).

Bijker, W. E., *Of Bicycles, Bakelites, and Bulbs: Toward a Theory of Sociotechnical Change* (Cambridge, MA: MIT Press, 1995).

Blok, A., M. Nakazora, and B. R. Winthereik, 'Infrastructuring Environments', *Science as Culture*, 25.1 (2016): 1–22.

Blok, A., and R. Tschötschel, 'World Port Cities as Cosmopolitan Risk Community: Mapping Urban Climate Policy Experiments in Europe and East Asia', *Environment and Planning D: Society & Space*, 34.4 (2016): 717–736.

Bolton, R., V. Lagendijk, V, and A. Silvast, 'Grand Visions and Pragmatic Integration: Exploring the Evolution of Europe's Electricity Regime', *Environmental Innovation and Societal Transitions*, 32 (2019): 55–68.

Boucher, A., B. Gaver, T. Kerridge, M. Michael, and others, *Energy Babble* (Manchester: Mattering Press,).

Bowker, G. C., and S. L. Star, *Sorting Things Out: Classification and its Consequences* (Cambridge, MA: MIT Press, 1999).

Boyer, D., *Energopolitics* (Durham, NC: Duke University Press, 2019).

Boyer, D., and I. Szeman, 'Breaking the Impasse: The Rise of Energy Humanities', *University Affairs*, March 2014: 40.

Burgess, J., and M. Nye, 'Rematerializing Energy Use Through Transparent Monitoring Systems', *Energy Policy*, 36 (2008): 4454–4459.

Buttny, R., and A. M. Feldpausch-Parker, 'Editorial: Communicating Hydrofracking', *Environmental Communication*, 3 (2016): 289–291.

Callon, M., 'The Sociology of an Actor-Network: The Case of the Electric Vehicle', in M. Callon, J. Law and A. Rip, eds, *Mapping the Dynamics of Science and Technology* (London: Palgrave Macmillan,1986), pp. 19–34.

Cartwright, E., 'Eco-Risk and the Case of Fracking', in S. Strauss, S. Rupp, and T. Love, eds, *Cultures of Energy: Power, Practices, Technologies* (Walnut Creek: Left Coast Press, 2013), pp. 201–212.

Chakrabarty, D., 'The Politics of Climate Change is More than the Politics of Capitalism', *Theory, Culture & Society*, 34.2–3 (2017): 25–37.

Cross, J., *Dream Zones: Anticipating Capitalism and Development in India* (London: Pluto Press, 2014).

de Rijke, K., 'Hydraulically fractured: Unconventional gas and anthropology', *Anthropology Today*, 29 (2013): 13–17.

Edwards, P. N., G. C. Bowker, S. J. Jackson, and R. Williams, 'Introduction: An Agenda for Infrastructure Studies', *Journal of the Association for Information Systems*, 10.5 (2009): 364–374.

Ernstoff, A., and B. R. Ellis, 'Clearing the Waters of the Fracking Debate', *Michigan Journal of Sustainability*, 1 (2013): 109–129.

Gabrys, J., 'A Cosmopolitics of Energy: Diverging Materialities and Hesitating Practices', *Environment and Planning A*, 46 (2014): 2095–2109.

Ghosh, A. 'The Imam and the Indian', in A. Bammer, ed., *Displacements: Cultural Identities in Question* (Bloomington: Indiana University Press, 1986), pp. 47–55.

Gupta, A., 'An Anthropology of Electricity from the Global South', *Cultural Anthropology*, 30.4 (2015): 555–568.

Gupta, A., and J. Ferguson, *Anthropological Locations: Boundaries and Grounds of a Field Science* (Berkeley: University of California Press, 1997).

Haraway, D., A Game of Cat's Cradle. Science Studies, Feminist Theory, Cultural Studies, *Configurations*, 1 (1994): 59–71.

— 'Anthropocene, Capitalocene, Plantationocene, Chthulucene: Making Kin', *Environmental Humanities*, 6 (2015): 159–165.

— *Staying with the Trouble: Making Kin in the Chthulucene* (Durham: Duke University Press, 2016).

Hargreaves, T., M. Nye, and J. Burgess, 'Making Energy Visible: A Qualitative Field Study of how Householders Interact with Feedback from Smart Energy Monitors', *Energy Policy*, 38 (2010): 6111–6119.

Harvey, P., C. B. Jensen, and A. Morita, *Infrastructures and Social Complexity: A Companion* (London: Routledge, 2017).

Holbraad, M., and M. A. Pedersen, *The Ontological Turn: An Anthropological Exposition* (Cambridge, UK: Cambridge University Press, 2017).

Howe, C., and D. Boyer, 'Aeolian Politics', *Distinktion: Scandinavian Journal of Social Theory*, 16.1 (2015a): 31–48.

— 'Wind at the Margins of the State: Autonomy and Renewable Energy Development in Southern Mexico', in J. McNeish, A. Borchgrevnik, and O. Logan, eds, *Contested Powers: The Politics of Energy and Development in Latin America* (Chicago: University of Chicago, 2015b), pp. 93–115.

Hymes, D., 'Ethnopoetics', *Theory, Culture & Society*, 23.2–3 (2006): 67–69.

Ingold, T. and E. Hallam, *Creativity and Cultural Improvisation* (London: Bloomsbury, 2008).

Jensen, C. B. 'Wound-up Worlds and The Wind-up Girl: On the Anthropology of Climate Change and Climate Fiction', *Tapuya: Latin American Science, Technology and Society* 1. 1 (2018): 50–64.

Latour, B., *Aramis, or, The Love of Technology* (Cambridge, MA: Harvard University Press, 1996).

Law, J., *After Method: Mess in Social Science Research* (Abingdon: Routledge, 2004).

Lepselter, S. 'The License: Poetics, Power, and the Uncanny', in D. Battaglia, ed., *E.T. Culture: Anthropology in Outerspaces* (Durham, NC: Duke University Press, 2006).

Maguire, J., 'The Temporal Politics of Anthropogenic Earthquakes: Acceleration, Anticipation, and Energy in Iceland', *Time and Society*, Online First (2019).

— 'Icelandic Geopower: Accelerating and Infrastructuring Energy Landscapes' (PhD Thesis, IT University of Copenhagen, 2017).

Maguire, J., and B. Winthereik, 'Living with the Earth: More-than-Human Arrangements in Seismic Landscapes', in P. Harvey, C. B. Jensen, and A. Morita, eds, *Infrastructures and Social Complexity: A Routledge Companion* (London: Routledge, 2017), pp. 161–173.

Matz, J., and D. Renfrew, 'Selling "Fracking": Energy in Depth and the Marcellus Shale', *Environmental Communication*, 9.3 (2015): 288–306.

McNeish, J. A., and O. Logan, *Flammable Societies: Studies on the Socio-Economics of Oil and Gas* (London: Pluto Press, 2012).

Mitchell, T., 'Carbon Democracy', *Economy and Society*, 38.3 (2009): 399–432.

Mol, A., *The Body Multiple: Ontology in Medical Practice* (Durham: Duke University Press, 2002).

Nader, L., *The Energy Reader* (Oxford: Wiley-Blackwell, 2010).

Pinkus, K., *Fuel: A Speculative Dictionary* (Minneapolis: University of Minnesota Press, 2016).

Raffles, H., *Insectopedia* (New York: Random House Inc, 2011).

Richardson, L., 'Poetics, Dramatics, and Transgressive Validity: The Case of the Skipped Line', *Sociological Quarterly*, 34.4 (1993): 695–710.

Schick, L., and B. R. Winthereik, 'Innovating Relations – or Why Smart Grid is not too Complex for the Public', *Science and Technology Studies*, 26.3 (2013): 82–102.

Shove, E., and G. Walker, 'What is Energy For? Social Practice and Energy Demand', *Theory, Culture & Society*, 31.5 (2014): 41–58.

Silvast, A., 'Energy, Economics, and Performativity: Reviewing Theoretical Advances in Social Studies of Markets and Energy', *Energy Research & Social Science*, 34 (2017a): 4–12.

— *Making Electricity Resilient: Risk and Security in a Liberalized Infrastructure* (London: Routledge, 2017b).

Silvast, A., H. Hänninen, and S. Hyysalo, 'Energy in Society: Energy Systems and Infrastructures in Society', *Science and Technology Studies*, 28.3 (2013): 3–13.

Sovacool, B. K., 'What Are We Doing Here? Analyzing Fifteen Years of Energy Scholarship and Proposing a Social Science Research Agenda', *Energy Research & Social Science*, 1 (2014), 1–29.

Star, S. L., and K. Ruhleder, 'Steps Toward an Ecology of Infrastructure: Design and Access for Large Information Spaces', *Information Systems Research*, 7.1 (1996): 111–134.

Strathern, M., *Partial Connections* (Lanham: Rowman and Littlefield Publishers, 1991).

Strauss, S., S. Rupp, S,and T. Love, *Cultures of Energy: Power, Practices, Technologies* (Walnut Creek, CA: Left Coast Press, 2013).

Szeman, I., 'What the Frack? Combustible Water and Other Late Capitalist Novelties', *Radical Philosophy*, 177 (2013): 2–7.

Szeman, I., L. Badia,.J. Diamanti, M. O'Driscoll, and M. Simpson, *After Oil* (Alberta: Petrocultures Research Group, 2016).

Szeman, I., and D. Boyer, *Energy Humanities: An Anthology* (Baltimore: JHU Press, 2017).

Szeman, I., Wenzel, J., and P. Yaeger, *Fueling Culture: 101 Words for Energy and Environment* (Oxford: Oxford University Press, 2017).

Tamagno, B., 'Iceland people and Environment Interact in the Land of Ice and Fire', *Geodate*, 27.1 (2014): 2.

Tsing, A., 'Worlding the Matsutake Diaspora, or can Actor–Network Theory Experiment with Holism?', in T. Otto and N. Bubandt, eds, *Experiments in Holism: Theory and Practice in Contemporary Anthropology* (Malden, MA: Blackwell Publishers, 2010), pp. 47–66.

Tsing, A., and E. Pollman, 'Global Futures: The Game', in D. Rosenberg and S. Harding, eds, *Histories of the Future* (Durham, NC: Duke University Press, 2005), pp. 107–122.

Urry, J., 'The Problem of Energy', *Theory, Culture & Society*, 31.5 (2014): 3–20.

Watts, L., *Energy at the End of the World. An Orkney Islands Saga* (Cambridge, Massachusetts: MIT Press, 2019).

Watts, L., and B. R. Winthereik, 'Ocean Energy at the Edge', in G. Wright, S. Kerr, and K. Johnson, eds, *Ocean Energy: Governance Challenges for Wave and Tidal Stream Technologies* (London: Routledge, 2018), pp. 229–246.

Wilkie, A., and M. Michael, 'Designing and Doing: Enacting Energy-and-Community', in N. Marres, M. Guggenheim, and A. Wilkie, eds, *Inventing the Social* (Manchester: Mattering Press, 2018), pp. 125–148.

Willow, A., and S. Wylie, 'Politics, Ecology, and the New Anthropology of Energy: Exploring the Emerging Frontiers of Hydraulic Fracking', *Journal of Political Ecology*, 21 (2014): 222–236.

Wilson, S., Carlson, A., and I. Szeman, *Petrocultures: Oil, Politics, Culture* (Montreal: McGill-Queens University Press, 2017).

Winther, T., and H. Wilhite, 'Tentacles of Modernity: Why Electricity Needs Anthropology', *Cultural Anthropology*, 30.4 (2015): 569–577.

Winthereik, B. R., J. Maguire, and L. Watts, 'The Energy Walk: Infrastructuring the Imagination', in D. Ribes and J. Vertesi, eds, *Handbook of Digital STS* (Princeton: Princeton University Press, 2019).

Winthereik, B., 'Is ANT's Radical Empiricism Ethnographic?', in A. Blok, I. Farias, and C. Roberts, eds, *The Routledge Companion to Actor Network Theory* (New York: Routledge, 2019), pp. 24–33.

Wylie, S., 'Fractivism: Corporate Bodies and Chemical Bonds, *Conservation & Society*, 16.4 (2018): 525–526.

2

THE POWER OF STORIES

Ann-Sofie Kall, Rebecca Ford and Lea Schick

STORIES HAVE POWER: POWER TO COMMUNICATE, GATHER, ENTERTAIN AND EDUCATE; power to seduce and enchant, convince and transform both people and societies. Energy infrastructures are filled with stories, stories that have many kinds of power. This chapter is about the different powers to be found in stories of infrastructures, from waste incinerators, to marine energy, to nuclear power. And it asks: What do these stories of energy infrastructure have the power to do?

In this chapter, we tell stories about energy infrastructures, and through these we investigate how energy and power are inevitably entangled in one another. Rather than telling one powerful story, we will do this by weaving together a collection of six short stories, each situated in a particular time and place. These stories demonstrate the diversity and versatility in energy infrastructures, as well as in storytelling practices.

In our narratives we travel to three countries and visit different field sites, each bound to the other through specific material objects. We travel from the Orkney Islands, off the north coast of Scotland, where the community is developing renewable technology to harness the power of wind, wave and tide; then we move to a waste-to-energy incinerator in central Copenhagen, Denmark, and finally to Stockholm, Sweden, to the archives and parliamentary documents concerning nuclear power that connect to the building of wind turbines in the province of Östergötland. We pay attention to how specific stories are made, and how some become important while others are silenced. We also investigate which stories travel, how they travel, how they gain (or lose) power and how they make things happen in the world. Through these six energy stories, we explore how local and situated stories enter into conversation with, as well as challenge and interweave, larger and global 'grand narratives' of energy.

Stories, like energy, have the power to make a difference in the world. Stories can travel and are themselves forms of energy transfer. They emerge from, and are woven into, the interactive process of meaning-making. Energy infrastructure stories, like all stories, live in the connections they make between people and places. In characterising this relationship between 'the work and the world' as a process of 'uninterrupted exchange', Mikhail Bakhtin likens it to the 'exchange of matter between living organisms and the environment' (Bakhtin 1981: 254). Similarly, Donna Haraway also emphasises the relationship between stories and the lived environment. For Haraway, stories 'propose and enact patterns for participants to inhabit'; they include 'science fiction, speculative feminism, science fantasy, speculative fabulation, science fact'. Because of the way stories are entangled with, and emerge through, living processes, they reflect the cultural situatedness of the worlds in which they are made, and so re-make that world. As Haraway points out, 'it matters what stories we tell to tell other stories with', just as, 'it matters what stories make worlds, what worlds make stories' (Haraway 2016: 10).

Stories are situated (Haraway 1988).[1] It matters which stories are being told, by whom, and in what situation. Not all stories are equal. Certain stories are not welcome in certain places, and stories that at one time seemed utopian/dystopian, or 'science fiction-like', may in time come to describe a new reality. As Jaber Gubrium and James Holstein write: 'Stories are assembled and told to someone, somewhere, at some time, for different purposes, and with a variety of consequences' (2009: 10).

Paying attention to these 'situated stories' can help us to understand the ways they are in dialogue with global narratives about energy futures and climate change. Anna Tsing describes the desire to create a 'global dream space' through appeals to 'universal truths', which can 'take us out of our imagined isolation into a space of unity and transcendence'. Yet in doing so 'we find ourselves, not everywhere, but somewhere in particular' (2005: 85). Tsing draws attention to the friction present in this process of situated storytelling, as global narratives are retold in different times and places. While the challenges of climate change and our future energy needs lead to calls for a global response, such a narrative unhelpfully suggests there can be a single universal solution. In attending to the specificity of our stories, and their friction, we the authors of this chapter hope to widen the imaginary space of opportunity in which future energy worlds may be told and made.

The chapter's six stories explore the storytelling process of specific energy infrastructures. By following these stories we see how they travel and act in the world. We zoom into the messy work of making energy, and into the different types of power it generates. We show the fragments of stories that do not make it into the official narratives, and their effects (and affects).

Stories of energy infrastructures are products of complex socio-technical and political negotiations (Jasanoff and Kim 2013; Stirling 2014; Sylvast et al. 2013). For example, the nation-state building work in the story of Sweden's nuclear power, or the tidal power in the sea around the UK. Energy infrastructure stories make claims about, and on, the world. They are given authority over energy systems and energy worlds. It therefore matters who gets to tell stories about whose energy infrastructures and their futures. Hopefully this chapter will ignite a spark in the reader's imagination to begin telling stories that have the power to spur the imagination of decision-makers for possible energy worlds beyond our current ones.

MARINE WASTE AND THE 'NARRATIVE OF DISAPPOINTMENT'

Rebecca Ford

A brief walk along any beach in Orkney shows you that there is more than seaweed and shells being washed up by wave and tide. Following the Orkney Beachcombing page on Facebook, I had become fascinated by the finds and the stories they told. I approached the page's author and asked if I could go out with him on a beachcombing expedition at Billia Croo, the European Marine Energy Centre (EMEC) wave test site, just outside Stromness. I'd walked on Orkney beaches often enough to have some idea of the amount of plastic that washes ashore. The unfamiliar writing on the labels of some of these plastic containers told me that they had come from abroad. The Orkney beachcomber could tell me where these things had come from, and how they might be reused.

A piece of orange plastic caught our eye, which the beachcomber immediately identified as a trap owner's tag from the Lobster Fishery in North America. The low licence number on this tag caused a bit of speculation, as other tags that come ashore tend to have four or five digits. Back home, the beachcomber posted the photo on an online lobster-fishing group, and the tag's owner got in touch from Tenants Harbour, Maine. Corey Morris thought the tag was lost 5–7 years ago, and explained that the license number was low because it had been that of his grandfather (who, back in the 1940s, was the 167th person to receive a lobstering licence); he had gifted his number to his grandson nearly 30 years ago, when Corey started lobstering.

Alongside the tag, and often entangled in nets and rope, floats and creel hooks, are many items less easy to link to their original owners. The degraded plastic bags, food wrappers, plastic bottles and sea-worn flip-flops are part of a vast quantity of anonymous rubbish which is difficult to place. I had never really pondered how so much stuff ends up in the ocean. The beachcomber helped me to understand how rubbish discarded on a street in

FIG. 2.1 Fishing tag at Billia Croo

a riverside city blows into the water and then, following that river down to the sea, is carried by currents, across oceans, to be washed up on faraway shores. The sea carries the consequences of our carelessness. The beachcomber taught me how to pay attention – how to read the stories left discarded on the shore.

Looking out across Stromness harbour one day, something caught my attention: a trailer sitting on the pier, loaded with massive cylinders. Their distinctive red and yellow paint made them immediately recognisable as part of a Pelamis wave energy device. I'd heard folk saying that one of the devices was being decommissioned – a less brutal way of saying chopped up and carted away for scrap.

Pelamis was one of the first devices to be tested at EMEC. Its image appeared not just on industry brochures and policy documents, but even in the Orkney tourist brochure – its long, jointed body earning it the nickname 'sea snake'. It had become part of the marine landscape of the islands. But here it was dissected and dismantled.

The fate of Pelamis can be explained by a downturn in the fortunes of the wave energy sector, which can be linked to what I call the 'narrative of disappointment'. This narrative emerged from a series of events revealing the power of stories, and illustrates the importance of how stories are told, and by whom.

In 2010 the Crown Estate, who manage the seabed around Orkney, suddenly announced the leasing of ten sites for commercial wave and tidal energy development. In the publicity

FIG. 2.2 Decommissioned Pelamis

surrounding this announcement they set developers the challenge of generating 1.2 GW of electricity by 2020 – the equivalent of the power needed by 750,000 homes. The Crown Estate's remit was to maximise income from its assets – at the time, this income went directly to the UK government.[2] The local community, whose coastal assets were effectively being enclosed and leased for profit, were not consulted; neither was the wave energy industry.

The reality was that the technology was far from ready; as I heard one person from the renewable energy sector put it, 'it was like asking the Wright brothers to build Concorde'. Generating 1.2GW by 2020 was never a realistic target; the feeling I have heard expressed is that the Crown Estate was encouraging developers to overestimate and provide best-case projections to attract investment. The narrative of disappointment characterises the failure to meet this target as a failure of the technology, leading to a decline in investor confidence. Without investment, developers did indeed fail, but for economic rather than technical reasons. In November 2014, Pelamis called in administrators after struggling with cash flow. Within a year another company, Aquamarine Power, developer of the Oyster wave device, was also in administration. EMEC's test site at Billia Croo had lost two of its longest resident devices.

The renewable sector in Orkney has created around 300 jobs and contributes to demand for other local businesses. In an island economy where employment options are limited and often poorly paid, part-time or seasonal, the sector opens up opportunities for young islanders to stay, or return to the islands, by providing skilled and graduate level jobs. Jobs or investment lost to the islands are hard to replace; when families face the difficult decision to look outside Orkney for work, the community loses more than a job, it loses part of itself, with repercussions for local schools, organisations and cultural activities. The Pelamis sea snake and Oyster had become part of the Orkney community; they were kin to those who worked with them and cared for them. The entanglement of technology into the process of community meant that the narrative of disappointment became woven into personal stories of redundancy, which in turn changed the shape of the community. When individuals and families left, their loss was felt from school classrooms to sports clubs and charity committees; you could trace their connections and contributions to the daily interactions and interdependencies we call community. None of this could be explained by a spreadsheet or a press release.

In Orkney, attempts to harness the power of the waves and tide to generate electricity have revealed the entanglement of nature and culture in this process of technology development – what Donna Haraway terms 'natureculture' (2016). The sea literally shapes the technology, as those who come with tank-tested prototypes soon discover; computer models don't contain large pieces of marine debris. At the same time, policy decisions and changes in the economic environment can suddenly make a technology unviable, regardless of its capabilities or potential.

The sea disrupts notions of fixedness; it defies a singular linear narrative. Ownership, authority and resources are constantly problematised and displaced, as tides and currents shift the contents of the seas around the globe. Like language, the sea is constantly in flux, shaping and being shaped by the things it comes into contact with. The waves show us the shape of things unseen – the strength of the wind, flow of the tide, depth of the water, texture of the seabed, the hidden reef. In the same way, the stories we tell reveal our world-view, our hidden concerns, the focus of our attention. What might we learn from this story of disappointment told by our marine waste? Are the powerful stories of competitive capitalism and economic growth helping us face the technological challenges of global warming, peak oil and plastic pollution?

I turn away from the abandoned remains of Pelamis and walk down to the shore. Here in the harbour the waves are small, gentle, domestic – how easy to imagine we have them tamed. At my feet a plastic bottle wrapped in seaweed catches my eye, another thing discarded when it ceased to be of use. Picking it up I automatically check inside for a message, but now I know even the empty ones have stories to tell. The label 'Egekilde Dansk Vand' tells me that it has come all the way from Denmark.

INTERMEZZO

One man's rubbish is another man's treasure. Let's follow this plastic bottle over the seas to Copenhagen. Here, plastic is a protagonist in a story about waste, energy, sustainability and tax money. A big ship full of Scottish rubbish has just arrived – it will be turned into district heating and electricity for the Danes.

WASTE WARS

Lea Schick

FIG. 2.3 Internal view of the 10th floor of Amager Bakke, Copenhagen

I'm standing on a metal footbridge suspended 60 metres over ground in the new waste-to-energy incinerator currently under construction in Copenhagen. Below my feet are the big ovens which will soon transform waste to energy. I'm being shown around by an employee responsible for the future visitor centre. Excited about this new and spectacular building he says:

> This is like standing on the 'elephant bridge' in 'Star Wars, The Force Awakens' where Kylo Ren [Darth Vader the second] is battling and killing his father, Han Solo. (Head of visitor centre at ARC, September 2016)

The energy produced in this room, however, is much more mundane and 'trashy' than swinging lightsabres and the good and evil forces of *Star Wars*. The war that is fought here concerns waste, energy, CO_2 and narratives about futures.

The new energy incinerator, named Amager Bakke (Amager is an area in Copenhagen, and Bakke means hill), is shaped like a mountain, and the public will be able to ski on its roof.[3] Amager Bakke replaces the old incinerator which each year transforms 440,000 tons of waste into district heating and electricity. In line with the global narrative of Denmark as a green frontrunner, the owner of the new incinerator, Amager Resource Centre (ARC), writes: 'with the establishment of Amager Bakke we are building one of the world's most environmentally effective incinerators, which will set new environmental standards in Denmark as well as internationally' [my translation].[4]

Even though the new incineration ovens are 20% more efficient, the new facilities are built with a much larger capacity than the existing ones. When I talk to the communications director at ARC, he tells me that, 'this is standard engineering practice – to build everything so that it is future proof' (i.e., larger) [my translation]. The question is, however, *which* future ARC is securing? This is a story of how the construction of infrastructures is a matter of making some futures come into being at the expense of others – a story about contested and conflicting futures (Brown et al. 2000; Jensen 2010; Adam and Groves 2007).

In 2010, when ARC proposed the plans for a new incinerator and requested the five municipalities owning the facilities to grant a loan guarantee of 4 billion DKr (€ 540 million), they encountered other and conflicting narratives about the future. The Danish Ministry of the Environment was in the midst of formulating a new 'Waste and Resource Strategy' with the subtitle '*Denmark without waste: recycle more – burn less*' (Miljøministeriet 2013). According to the Minister of the Environment at the time, Ida Auken, it would not be 'future proof' to build a new incinerator hungry for large amounts of waste – there would simply be less rubbish to burn. For Auken, waste incineration was a technology belonging to the past. Furthermore, Copenhagen Municipality had just agreed ambitious plans to become the world's first CO_2-neutral capital by 2025 (Københavns Kommune 2012). For this goal, the burning of waste – especially plastic rubbish, which emits more CO_2 than average household waste – had to be reduced immensely, and the municipality had therefore launched several initiatives to increase recycling. Based on these new trajectories of waste management, Copenhagen Municipality decided to reject the loan warranty. The five municipalities requested ARC to make a new tender bid for a smaller and cheaper incinerator. ARC, however, chose to ignore this, because an EU tender bid had already been released, and making a new one would delay the process and be 'a waste of money', as the former CEO of ARC, Ulla Röttger, said on Danish national TV (DR 2017).

In January 2011, ARC publicly announced that the architect Bjarke Ingels and his company 'BIG' (Bjarke Ingels Group) had won the architecture competition, and the compelling pictures quickly travelled the world's media, which praised his ideas of 'hedonistic sustainability' (more on this in the fourth story). And even though the funding had not yet been procured, in January 2012 ARC signed an order worth 1 billion DKr. with the technology provider B&W Vølund, who were to deliver the two new ovens (Pedersen 2012). Around the same time, the company changed its name from Amager Waste Incineration to Amager Resource Centre. The company, as well as 'waste', was rebranded as a resource for society (Habermann 2013).

The project was put on hold while secretive negotiations took place between municipality politicians and the government. Exactly what happened during these negotiations is not clear, but Vølund publicly announced that it had tried to influence the Minister of Finance, Bjarne Corydon (Ellers 2012; Hegelund and Mose 2013). Ida Auken later declared that she felt blackmailed to write a letter, together with Corydon, to the municipal politicians declaring that the incinerator would not be against the government's new resource strategy (Martini and Sandøe 2016; Hegelund and Mose 2013). In September 2012, the five municipalities permitted the loan guarantee with the restrictions that ARC could not use its full capacity and never burn more than the current amount of waste. ARC was not allowed to import waste from outside the five municipalities (DR 2017).

In 2013 the construction finally began. However, while the 'mountain' grew, its reputation crumbled. The whole process was widely criticised. The media generally agreed that the process was very controversial, and that ARC had lobbied and used the prospect of many jobs to persuade politicians to comply, though the business case was not viable (Bredsdorff and Wittrup 2015a; Martini and Sandøe 2016; Hegelund and Mose 2013).

FIG. 2.4 Amager Bakke, under construction, 2016

Only a few years later the critics were proved right. Due to decreasing amounts of waste as well as decreasing energy prices, ARC was projected to run a deficit of 1.9 billion DKr. over the next 20 years (Bredsdorff and Wittrup 2015b). As the TV documentary *Waste of Your Money* explicates, this situation put the five municipalities in an ambivalent position: either taxpayers would have to pay for the deficit through higher energy prices, or the municipalities would have to allow ARC to import waste and produce more energy, thus compromising Copenhagen's climate ambitions (DR 2017).

The politicians decided to allow the import of waste from other countries, such as Scotland and Sweden. Thus, Denmark may end up as 'the garbage can of Europe', as opponents had warned (Bredsdorff 2013). Maybe Hamlet was far-sighted when he said 'something is rotten in the state of Denmark' (Shakespeare 1603/2009). The import and burning of waste, however, raises questions around the borders of sustainability and CO_2 emissions. While the burning of imported waste may increase the local level of CO_2 emissions in Copenhagen, burning waste in 'the world's most environmentally effective incinerators' is still much better for the global environment than depositing waste in landfills or burning it in old incinerators.[5] The metropolitan competition to become the most sustainable city may be a forceful driver of sustainable transitions, while at the same time leading to less sustainable futures on a global scale because, as showcased in Copenhagen, becoming the world's first CO_2-neutral capital may be seen as more important than bringing down global CO_2 emissions. In the end the municipality decided to import and burn international waste and thus challenge their own CO_2 goals.

In Denmark, Amager Bakke has become an incinerator – and within politics and the public media it has become an appropriately hot topic. Most media stories define *Amager Bakke* as a complete failure, and there is widespread consensus that it is indeed a 'waste of taxpayers' money' (DR 2017). Building a larger capacity waste-to-energy incinerator was intended as a 'future-proof' investment – but the future changed! Which future ARC is 'proofing' – or maybe 'creating' – is not contained within a single narrative. Whereas opponents argue that ARC is causing unsustainability in Copenhagen, and that the money would have been better invested in recycling technologies, ARC continues to see itself as an important actor in the quest for sustainability. Its more efficient and less polluting ovens are being exported as 'green tech' to countries such as England and China, where they are replacing either landfills or coal power (Grundtvig 2017). Not all waste can be recycled, and ARC wants to develop a new market for Refuse Derived Fuel (RDF) – EU language for fuel made from waste from which all recyclable and harmful material has been removed (European Union 2003; Grundtvig 2017). It's better for the environment to make this into energy, thereby replace the burning of coal, than it is to let it rot in landfills. Building a new market which prices RDF higher than non-sorted waste may incentivise more recycling. Whether the burning of waste is 'sustainable' or not is not a straightforward question. If the energy produced replaces wind energy, it may not be as 'green', but if this waste energy is used as backup, instead of coal, in periods when it isn't windy, it may count as 'sustainable'.

From this 'waste war' in Copenhagen emerges other stories of 'future wars' and 'climate wars'. The 'forces' of 'good and evil' are not as clear-cut as in *Star Wars*. They are narrative forms up for negotiation. As Amager Bakke's director of communications tells me:

> The mind-set around waste incineration has changed immensely over the past decade since the plans for Amager Bakke began. Earlier no-one cared about waste. Now Amager Bakke has become a megaphone for all sorts of opinions around waste incineration [my translation].

What we can see from this story is that the building of new energy infrastructures is surrounded by and entwined in multiple narratives that don't stay still. As politicians and utility companies negotiate and plan new infrastructure developments, they take these emerging narratives into account. However, what this analysis shows is that such narratives are situated in, and account for, just one version of events at a time. As infrastructures-in-the-making emerge into a world filled with shifting constellations of publics and political plans, it's clear that planning for more sustainable energy futures is complex. What this analysis (itself a situated story) wants to add is that the temporality of narratives – their variability, periodicity, and multiplicity – not only emerge through and weave into the discursive fabric of actors' accounts, but also become the grounds on which the legitimacy of waste wars and climate futures are ambiguously negotiated and contested.

INTERMEZZO

What counts as good and evil energy is situated in particular socio-political environments and may change over time. Thanks to strong public resistance in the 1970s, nuclear power is seen as the dark side of energy and Denmark does not have any nuclear powerplants. What most Danes do not know, however, is that 3% of the electrons powering their homes are produced in Swedish nuclear powerplants. Through a thick power cable under the Kattegat Sea we will now follow these electrons back to the Swedish Parliament where they are negotiated.

NUCLEAR STORIES

Ann-Sofie Kall

I'm sitting in the archive, surrounded by old boxes, piles of papers, handwritten letters, blueprints, notes, and sometimes even pressed flowers and poems. I am following threads, linking events and finding stories.

It is Birgitta Hambraeus's personal archive in Stockholm, part of The Swedish National Archives. At the beginning of the seventies Birgitta was a member of the Swedish Parliament, known for bringing resistance to nuclear power onto the parliamentary agenda. Together with physicist and Nobel Prize winner Hannes Alfvén, she has been described as the most important actor for the breakthrough of nuclear critique in Sweden (Anshelm 2000: 119). Their action paved the way for a change of direction no one predicted at the time. At the beginning of 1972, nuclear power was considered the rational choice and had been the preferred energy solution since the late 1950s (Lindquist 1997). A year later, nuclear power was the subject of lively discussions, both in parliament and in public debate. In 1971, Hambraeus was tasked by her party (Centerpartiet) with investigating whether peaceful

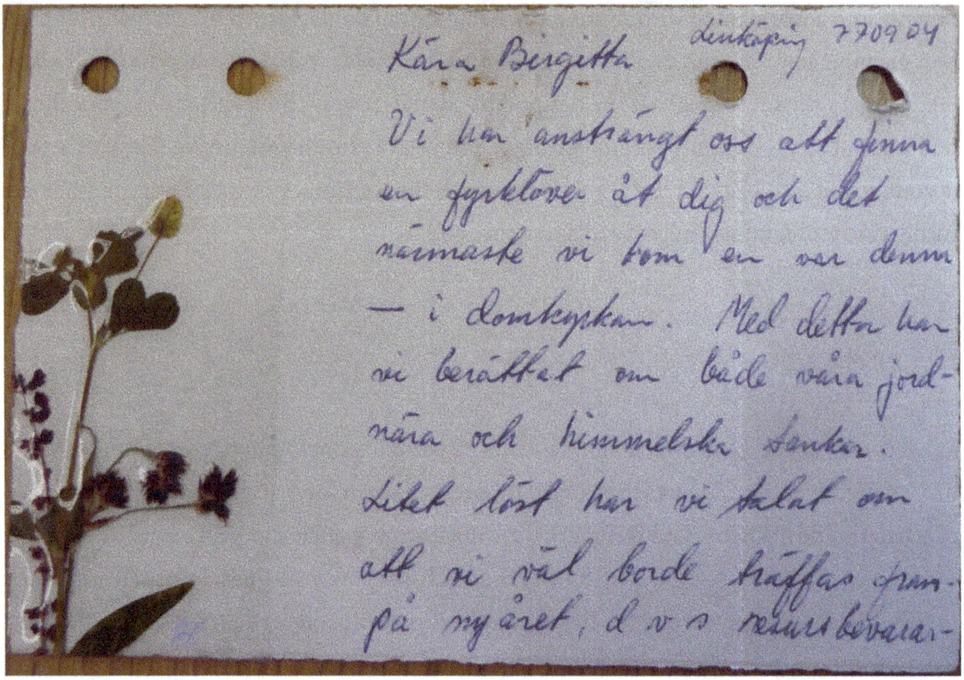
Kära Birgitta Linköping 770904
Vi har ansträngt oss att finna en fyrklöver åt dig och det närmaste vi kom en var denn — i domkyrkan. Med detta har vi berättat om både våra jordnära och himmelska tankar. Litet löst har vi talat om att vi väl borde träffas fram-på nyåret, d v s resursbevarar-

FIG. 2.5 Card sent to Birgitta Hambraeus

nuclear power could have a negative effect on people, society and the environment. In the general story of Hambraeus's achievements, it's described how she, with the help of experts and by using the instruments of parliamentary bureaucracy, succeeded in making nuclear power into a matter of concern.

Hambraeus placed great importance on creating networks and relations outside the Parliament, with scientists, activists and representatives of industry and the media, and with private individuals. Every morning when she came to work in Parliament, she went to her mailbox and got the day's pile of letters. She then used the rest of the day to read and write letters. This was the strategy she used to learn more about nuclear power, to create awareness and draw others to the issue, and to make new alliances by introducing different experts to each other.

I'm reading these letters, putting fragments together.

In one of the old boxes I find a letter that captures my interest. It is written by Birgitta Hambraeus and addressed to the author and poet Harry Martinson on 10 October 1972. He was a known writer, a member of the Swedish Academy, and received the Nobel Prize for Literature in 1974. Here is that letter:

10/10/1972

PhD. Harry Martinson

Swedish Academy

Stortorget

Stockholm

In my work with energy issues in Parliament I have discovered a problem of such enormous dimensions, I feel unable to express it so that it can be understood in its entire reach.

I now turn to the author of Aniara, and hope you will feel it is your task, with your artistic capacity, to make us realise what we are about to do.

Enclosed is the interpellation on the expansion of nuclear power I directed to the Minister of Industry Rune Johansson in Parliament.[6] I will get an answer from him in the Parliament chamber on November, 30th.

With respectful greeting

Birgitta Hambraeus

[my translation]

The letter was sent during the autumn of 1972, a time when Birgitta Hambraeus had an important role in the process of changing the direction of the Swedish debate on nuclear power and, not least, the politics in Parliament.

Aniara, the text mentioned in the letter, is a collection of 103 poems and a story about a large spaceship (Martinson 1956). It can be seen as a comment on environmental development. Everything starts with a routine mission in space, transporting 8,000 people to Mars due to nuclear war on Earth. As a result of an asteroid strike the spaceship loses its bearings and manoeuvrability. It is now hurtling out of our solar system and into the unknown. As a poem that is known for its critique of technology and as a symbol of our (human and the Earth's) vulnerability, it's no coincidence *Aniara* is mentioned in the letter.

The letter Birgitta Hambraeus sent to Harry Martinson from her position as Member of Parliament can be seen as an attempt to involve other stories and voices in the political debate about nuclear power. In analysing the history of Swedish parliamentary energy politics, I have learned that some aspects of energy politics are taken for granted, while others are hidden. Categories are created and distinctions are made between what is and is not relevant. Various realities are enacted that allow or prevent certain types of political action (Moser 2008). By engaging the author Harry Martinson and his work, Birgitta Hambraeus challenged the narrative dominating the way we talk about and act upon energy, not least within politics.

The energy system often becomes a matter of numbers and technical objects, and is reduced to something static and manageable. A clear distinction is maintained between what is described as objective, scientific and unambiguous, in contrast with what is being presented as non-scientific, political, ideological or irrational. Advocates of renewable energy and/or small-scale alternatives have been accused more than others of not acknowledging reality and of being too naïve and ideological (Kall 2011). Considering how entangled energy is in our everyday lives, it may seem strange that it could be reduced solely to terawatt-hours and technical devices. A letter to a poet and author, of course, will not solve humanity's problematic relations with nature, or lead to energy solutions without adverse effects. It can, however, open up more diverse stories, other perspectives, different practices and actors.

FIG. 2.6 Meeting room in the Swedish Parliament

INTERMEZZO

Stepping out of the archive it may seem that the stories from that period do not matter. Sweden still has its nuclear power, affecting people nearby and in neighbouring countries. The energy system is not yet transformed. However, renewable energy sources today comprise a substantial part of Sweden's energy supply. And the nuclear power plant Barsebäck, visible from Copenhagen in Denmark, where we now are heading, has been decommissioned.

NARRATIVES OF PROMISCUOUS, PROMETHEAN AND CAUTIOUS ARCHITECTURE

Lea Schick

> *Architecture is the canvas for the stories of our lives.*
>
> (BJARKE INGELS 2015)

It's a clear autumn day. I'm standing on the top of Amager Bakke. Just across the water dividing Denmark and Sweden I see two big grey cubes, recognisable as the former nuclear powerplant Barsebäck. This piece of architecture tells the story of a future that never came to pass in Denmark. According to Danish media it was thanks to strong Danish political opposition that the Barsebäck nuclear powerplant was closed down in 2005 (Dagbladet Information 2005).

The piece of architecture I'm standing on also tells stories – stories about sustainability and hedonism. Amager Bakke is an 85m man-made 'mountain' of waste. Its roof has been turned into a public park with trees, hiking-trails, artificial snow, ski-lifts and -slopes. The sides of the building are covered with vertical gardens and mountain-climbing trails imitating nothing less than routes on Mount Everest and Mont Blanc. With this architectural design ARC wants to tell new stories about waste, energy and sustainable futures. Over the past centuries many forms of powerplants (especially nuclear powerplants like Barsebäck) have been increasingly removed from residential areas, just as energy has been made invisible to the general public (Edwards 2003; Larkin 2013). ARC wants to change this by opening up the new building for the public to use as a place for recreation and entertainment: 'we want to be more than a powerplant – we want to become part of the city'.

FIG. 2.7 Amager Bakke

Amager Bakke was designed by the world-famous Bjarke Ingels Group (BIG architects). Rather than hiding away drab industrial processes like energy production, Ingels is designing what he calls 'social infrastructure'. His philosophy is that infrastructures should be turned into spaces of recreation where the public can enjoy themselves and where new kinds of public engagement with energy and sustainability have the potential to emerge (BIG Vortex 2012). With his architectural stories Ingels wants to make 'BIG' changes, such as redesigning our relations with infrastructures, architecture and sustainability. Thus, Amager Bakke should 'completely alter people's perception of a powerplant from that of a dirty neighbour to that of a public park'. Ingels (2012) describes Amager Bakke as a 'promiscuous hybrid', 'this idea of taking seemingly mutually exclusive ideas – that of a park and a powerplant and turning them into a power-park'. But the question becomes: how is power made visible on the ski slopes? How does the 'power-park' work as a technology for public engagement in energy and waste infrastructures?

During an interview, I ask the ARC communications director how electricity and waste will be visible to the public. He answers that this is not something that ARC will have much influence on, because the final design of the park, as well as its operations, will be outsourced to a third-party company. According to Danish law, a public utility is not allowed to profit from leisure-making activities. Like me, the ARC employee is concerned about the final design and the public's perception of the power-park. His concern is not so much with sustainability and the making of engaged energy publics, but with making particular sports publics: 'I don't like that the media and public describe Amager Bakke as an amusement park. It's important for me that this does not become pure entertainment. It needs to be a place for serious sports activities.'[7] Even though the intention is to change public and infrastructural relations, it's uncertain how waste, energy and sustainability will become a matter of concern for the public (Latour 2005).

The public engagement that Bjarke Ingels is designing for is a public engaged in sustainability through pleasure, joy and fun. He has labelled his architecture style 'hedonistic sustainability':

> Sustainability has always been associated with some form of sacrifice. [...] That you would accept downgrading your quality of life in order to become sustainable. But we were thinking, why not actually make a sustainable city or a sustainable building or a sustainable life even more enjoyable than the alternative?[8]

FIG. 2.8 Ski slope, Amager Bakke

For Ingels, cigars and smoke rings are the ultimate symbols of a hedonistic lifestyle (Schröder 2017).[9] His proposal for the BIG powerplant included a chimney which would emit large circular smoke rings that would float over Copenhagen.[10] As a form of collective eco-feedback technology, each smoke ring visualises the emission of one ton of CO_2. The smoke rings bear a striking resemblance to the art installation *Nuage Vert* by HeHe in Helsinki in 2008 (HeHe 2008). But whereas HeHe wants the artwork to provoke critical reflection on our current cornucopian lifestyle, Ingels' hedonistic sustainability tells a more promiscuous narrative. Rather than adopt a modest and humble lifestyle, we can blow smoke rings through the infinite cornucopian horn (chimney). Ingels' ideas, his architecture and his ostentatious vocabulary have travelled globally and gained the attention of publics, politicians and professionals. According to *TIME* magazine he was among the 100 most influential people in the world in 2016 (Koolhaas 2016). His emphasis on green architecture, new relations to infrastructure and not least on (hedonistic) sustainability is said to have had considerable impact on the global agenda on sustainable transitions. As the name of his company, and its slogan 'Yes is more' indicates Ingels (2009) is not exactly a humble man.

Bruno Latour shares Bjarke Ingels' concern for redesigning our world in a more ecological and sustainable way, and they both see architecture and infrastructure development as powerful tools for (re)imagining and remaking our reality – as tools for 'world-crafting' – 'the craft of making our world' (Ingels 2012). Latour (2008) says that design and architecture are redesigning our being-in-the-world. Ingels (2012) writes that architecture 'enables us to turn surreal dreams into inhabitable space. To turn fiction into fact'. But what kind of facts and what kind of futures does Amager Bakke assemble or design?

Ingels' architecture style and his promiscuous hybrids emerge as somewhat of an antagonist to Latour's *Cautious Prometheus* and his call for a more modest and humble way of redesigning the world.[11] The question is, what happens when a 'cautious Prometheus' meets Ingels' 'promiscuous hybrids' and 'hedonistic sustainability' – can these creatures mate and coexist in one building or architecture style? Would a cautious Prometheus be able to travel into TED talks and gain the same kind of political influence as Ingels' promiscuous hybrids and hedonistic sustainability? What would Ingels' architecture look like if it adopted some of the precautionary principles of the cautious Prometheus? That is, can we make new hybrid and sustainable energy infrastructures which are both promiscuous, and Promethean and cautious?

INTERMEZZO

Leaving these questions unanswered, our journey now takes us back to the Orkney Islands, where a people descended from the Danish Vikings are innovating in a much more modest and cautious way.

'ORKNEY'S ELECTRIC FUTURE'

Rebecca Ford

Having previously been subject to first the Norwegian and then the Danish crowns, in 1468 Orkney was pledged as a dowry by Christian I, King of Denmark, Norway and Sweden when his daughter Margaret married King James III of Scotland. The Norse influence is still evident in the islands, going beyond archaeological remains and place names to include cultural attitudes and questions of identity (Lange 2007). Just as the Nordic concept of the Law of Jante discourages individual success in favour of collective effort, within Orkney there is a dominant shared narrative about avoiding being 'bigsy', the dialect word for conceited, thinking you are better than other people and behaving in an arrogant or boastful manner (Flaws and Lamb 2001). Avoiding bigsy-ness has a powerful influence in shaping attitudes and communication within the islands (Ford 2013). It is related to another shared narrative – about the existence of a cohesive Orkney community, founded on ideals of egalitarianism and self-reliance through shared effort and cooperation (Matarasso 2012). This story persists in the face of the reality of island life, which of course includes inequality, self-interest, disagreement and division.

However, the story is persuasive enough to shape behaviour; people do cooperate and act as if there were such a thing as this ideal, egalitarian Orkney community. Walking through Orkney's capital, Kirkwall, I often see 'Orkney's Electric Future' printed on the side of one of the local authority's electric cars and bus. For Orkney Islands Council (OIC) this is the title of an ongoing project to promote electric vehicles and sustainable transport, making use of Orkney's abundant renewable energy to reduce transport costs and vehicle emissions. As I will tell it here, 'Orkney's Electric Future' is the title of a story, one that weaves together the various responses within the islands to narratives about energy, climate change and the politics of place.

FIG. 2.9 Orkney Islands Council electric vehicle

The 'narrative of disappointment', which emerges as part of the story of Orkney's marine renewable energy sector, is a story shaped by policy-makers and disillusioned investors whose relationship with the islands was based on economic opportunity. The response within the islands to the narrative of disappointment has grown out of ongoing entanglements and interdependencies; it emerges from particular, situated realities in dialogue with shared cultural narratives. It is woven into a meshwork of living relationships and interactions which bring marine plastic waste and decommissioned wave energy devices into the shared story of Orkney as an egalitarian community that values cooperation and cohesion; this is the story of Orkney's Electric Future.[12]

In 2016 Orkney produced 120.5% of its electricity demand, but the electricity grid was not designed for this generation (Energy of Orkney, 2017). The interconnector between Orkney and the Scottish mainland does not have the capacity to transport all of those Orkney electrons off the islands (Watts 2011). It gets so hot, in fact, that it's in danger of melting. In 2012 the network operator in Orkney, Scottish and Southern Electricity (SSE), called for a moratorium on new generation connections. To avoid overloading the interconnector during periods of peak generation it implemented a Smart Grid approach, using Active Network Management technology to monitor real-time production.

While this allows the grid infrastructure to operate safely within the limits of the interconnector, it also means some bigger renewable devices must be curtailed during periods of high production. Often it's the community-owned wind turbines that are switched off, losing both revenue and feed-in tariffs for the community money which is needed to pay back their investment and provide funding for local development projects.

The impetus for investing in turbines in the first place arises from a need to secure additional income to keep fragile island communities viable. The job situation on smaller islands is even more precarious than on Mainland Orkney: housing is scarce and often of poor quality, transport links are limited and weather dependent, everything costs more and takes longer to source. It's perhaps unsurprising that the age profile of these island populations is trending upwards while overall numbers decline, yet these communities, too, are committed to securing their future, and so are prepared to invest time and money to address the challenges of island living.

In a profit-driven electricity market, someone has to pay to upgrade the interconnector. Market regulations set the terms on which SSE can pass on the cost to consumers, and an economic case must be made to justify the expense of infrastructure investment. This is where the 'narrative of disappointment' has been woven into the story of 'Orkney's Electric Future'. Rather than waiting for an external solution to grid constraint, new schemes are being developed to store, and use, Orkney's abundance of renewable energy within the islands.

The Orkney 'Surf n' Turf' project received £1.5 million in funding from the Scottish Government's Local Energy Challenge Fund and created a partnership between Community Energy Scotland, European Marine Energy Centre (EMEC) and the Eday community wind turbine. Electricity generated by the community turbine and EMEC's tidal test site will be used to produce hydrogen, which will be shipped to Kirkwall, where a hydrogen fuel cell will provide electrical power for the island ferries when they are tied up at the pier overnight.

The potential to use hydrogen as an energy storage medium has attracted interest from outside the islands, too. In 2016 Orkney was chosen as the location for the BIG HIT – a five-year €5 million, EU project. An additional electrolyser will be installed on the island of Shapinsay, where the community wind turbine will be used to produce hydrogen. The scheme will fund ten electric vans for local authority use, fitted with hydrogen fuel cell range extenders, and the construction of a hydrogen refuelling station; in addition, there are plans to install two hydrogen-powered boilers in public buildings.

When I visited Eday to see the Surf n' Turf project I met members of Eday Renewable Energy (ERE), who operate the community wind turbine. They introduced me to Ernie Miller, who lives just along the road from the test site. Ernie sold some of his land to EMEC, to enable the Centre to build the tidal test site, and another plot to the community development trust to build their wind turbine. As a lifelong resident of the island he can remember a past before the electricity grid, when domestic-scale wind turbines were a common sight on crofts around Orkney. He is enthusiastic about the new generation of wind turbines, and the tidal and hydrogen technology shaping the future of the island. A wonderful photograph shows a smiling Ernie standing beside his vintage Lucas Freelite wind turbine, which he has restored to immaculate working condition – behind him is the Eday community turbine which will soon make the world's first wind- and tide-generated hydrogen.

Just as Ernie saw the mutual benefit of selling his land for renewable developments, EMEC and Eday Renewable Energy understood the importance of working together to overcome the grid constraints, which threatened to curtail all their efforts at energy production. There is a clear understanding of what is at stake: for EMEC the ability to continue to offer developers the grid-connected test facilities they need and thus maintain its leading role in the Marine Renewable Energy sector; for ERE, the ability to repay its investment and secure income to support the future of the Eday community. Alongside the celebratory headline of 120% renewable production sits the uncomfortable fact that Orkney has the highest level of fuel poverty in Scotland, with 63% of households spending more than 10% of their income on fuel costs. Low incomes, poor housing and higher transport costs for fuel are all contributing factors, as is the fact that Orkney's electricity costs 2p per unit more than other areas due to regional distribution charges.[13]

In Eday, Orkney's Electric Future is being made through a story of collaboration and cooperation between individuals, community groups, private business and local government. It's a story that weaves historical and cultural narratives into a pragmatic response to technical, geographical and social challenges. In attending to the particularities of the local context and the needs and concerns of the community, this story is imagining new ways of making, storing and using renewable energy, showing how care for people and place has become part of the 'living technology' vital for building liveable futures, in Orkney and elsewhere.[14]

INTERMEZZO

The wind crossing the North Sea to Sweden provides the power for another individual who is taking renewable energy production into his own hands.

FIG. 2.10 Ernie Miller, Eday

JOHNNY'S WIND TURBINE

Ann-Sofie Kall

Far from Orkney, in Östergötland in Sweden, where the landscape is flat and the terrain open, Johnny is building a wind turbine. This is the third model he has made, where all the experience and knowledge from the previous two are put into action. However, there are still some details left before this wind turbine is completed.

During the 1970s Johnny started to build his first turbine, a savonius rotor, which can be best described as a long-cut oil barrel. When he started his project the question of energy was one that provoked many weary, tough and lively political discussions in Sweden (Lindquist 1997; Anshelm 2000; Larsson 1986). This was also the time when the environmental movement, as in many Western countries, strongly influenced the energy debate. Modernity as an ideal, in which technology, science, development and economic growth had prominent roles, was no longer deemed self-evident, and the growth-oriented approach began to be replaced by a more ecological perspective. In connection with this critique of modernity, a contrasting vision of a different kind of society was developed. Two contrary narratives were created, building their own networks and formulating their own questions and solutions:

on the one hand, a centralised, large-scale, scientific and technical growth-based society; on the other hand, a more decentralised, small-scale, and locally-based low-energy society (Anshelm 2000; Wiklund 2006).

With a background working as an electrician at sea, Johnny was used to thinking in terms of small-scale and decentralised systems. He often spent weeks at a time offshore. Great knowledge of how different parts were connected was necessary. Solving problems, rebuilding what was already available, and finding new solutions when resources were limited, was central. For Johnny, living in one of the inland areas in Sweden with the best possible conditions for wind power, the idea of becoming a power producer and building his own wind turbine came naturally.

Since the seventies, renewable energy sources have served as a basis for both utopian and dystopian stories of the future, becoming representatives and symbols of certain types of society. They can be seen as ways to advance the small and the small-scale, to safeguard against regression and the anti-modern, and to uphold development and modernisation. They can also be seen as a sign of decline, a threat to development and the comfortable way of life (Kall 2011; Laird 2001).

FIG. 2.11 The wind turbine in Johnny's garden

Several things affected Johnny's thoughts about what energy is and how we actually use it. He believes that there is a lack of knowledge and understanding about how energy works and how much energy it takes to do everyday things, such as boiling eggs. According to Johnny, the distance between people and energy is far too great.

This tallies with research in the fields of energy systems and energy users, showing that there is a lack of knowledge about energy processes. For some, electricity is only about pressing a button on the wall. There is also a big gap between the producers and the consumers, situating households at the periphery of the technical system. Many lack knowledge about the technology, their own energy use and what draws a lot of power (Bladh 2007). It also works the other way around: many producers and innovators lack knowledge about the practices and perceptions of users (Rohracher 2003; Shove and Walker 2010). A deeper understanding is rare. This knowledge is something Johnny values highly and sees as important in understanding our relation to both resources and energy. For him, the construction of his wind turbines has been in many ways an education project. He is motivated by the challenge of learning more about what is possible and what is not. It has also been his goal to only use things that others do not need.

Standing next to the turbine, it's fascinating to hear Johnny telling stories of where its different parts come from. It also provides an opportunity to understand the potential of not buying new things or using ready-made solutions. For example, Johnny has used a T gear from a lawn mower, a rear axle from a Volvo truck and an angular gear from an aeroplane to build his wind turbine. With contacts and plenty of time, it has been possible to realise the project on a very low budget. From an economic point of view, however, it would probably have been more 'rational' to buy things new and thus connect

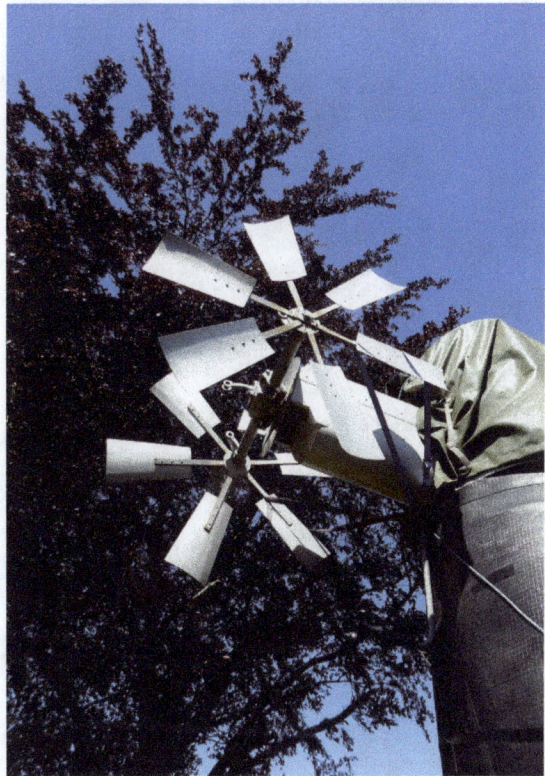

FIG. 2.12 The rotor blades of the wind turbine

to the electricity grid at an earlier stage. But for Johnny, it has been important to use what others have thrown away, or rebuild what was intended for something else. He is worried about how nature is being drained of resources without a thought for the consequences. Therefore, he wants to try to do something else. Attempting to convince others, however, has not been a driving force or a reason to do what he does.

Johnny's interest in energy issues is in many ways linked to resources, waste, renewables, and small-scale and decentralised energy – but it's difficult to place him in ready-made categories of, for example, proponents of renewable energy sources, or enthusiasts for small-scale technology. For Johnny, it's not necessarily that renewable energy sources or eco labelling are always preferable, it just does not make sense to move things over large distances, throw things away or always buy things new. He wants to learn everything there is to know about small-scale wind turbines, and he is happy to tell me all about it.

FIG. 2.13 Johnny in front of the control panel

It's difficult to portray Johnny as a role model to be followed by others. For most, building their own wind turbine is not an option. The interest, knowledge and devotion required to start, and stay, with such a project requires more effort than most are prepared to put in. The question is – what we can learn from this story and all the others, even if we do not want to weld our own wind turbine together? This is an example of yet another energy story, different from all the others, with its own meaning, actors, and material objects, showing the power of and in energy, and the power of people and stories.

CONCLUDING REFLECTIONS

It's never possible to tell the whole story, even though global narratives on energy sometimes give that impression. Through the specific located practices that we have studied, we experience how energy futures are (more or less) carefully constructed and how they are conflicted. Energy futures and energy worlds are being woven constantly through stories and practices in action. Stories, as well as energy infrastructures, are 'world-crafting', to borrow a term from one of our informants.

Writing together and combining various energy field sites is an experiment in stitching and weaving other kinds of energy worlds (Lindstöm and Ståhl 2014). Rather than a neatly crafted, coherent and strong narrative we have created a patchwork, or rather a meshwork, of multiple stories and their connections. In stitching the field sites together, we draw attention to the ways in which different places, materials, people and energies are in dialogue with one another. As Tim Ingold points out: 'the things of this world are their stories, identified not by fixed attributes but by their paths of movement in an unfolding field of relations' (Ingold 2011: 160).

Our style of storytelling is therefore deliberately fragmented and contains many voices – we do not attempt to draw out one coherent story or finding, but want the stories to remain open to this 'unfolding field of relations'. There are endless stories to be told about energy and the environment, climate change and fossil fuels, the promise of renewable technology and the dilemma over nuclear power. If we are to meet the energy challenges of the future, while also paying attention to relationships of power, these stories will be more important than ever.

NOTES

1 Following on from situated knowledges (Haraway 1988).

2 As part of the 2016 Scotland Act, Crown Estate Scotland was set up to allow the devolved Scottish Parliament to manage and receive the benefits from Crown Lands in Scotland. This was one of the recommendations of the Smith Commission, established following the 2014 Scottish Independence Referendum.

3 Amager Bakke opened in late 2019.

4 See www.a-r-c.dk.

5 Around 30% of Europe's waste is currently deposited in landfills.

6 An interpellation is a written question from a Member of Parliament to a government minister.

7 See Marres 2012 for an explanation of 'engaged publics'.

8 See Franklin-Wallis 2016 for Ingels' quote.

9 See Schröder 2017 for Ingels' quote.

10 Even though the smoke ring pictures are still present on ARC's homepage (www.a-r-c.dk), this idea has been dropped because it would be too energy intensive to create the smoke rings.

11 Bruno Latour uses the figure of the Greek god Prometheus, who stole fire from the gods and gave it to humans, as an analogy for the way humans (in the Global North) over the past centuries have consumed energy and other resources at an exponential pace, causing radical, and often fatal, environmental changes to the planet. Redesigning for a different way of inhabiting the planet, and co-existing with it, needs to be done with the same speed and energy as 'stealing fire from the sun'. However, it needs to be done much more carefully, considering how the stuff we make has a range of environmental effects in a variety of complex ways. This demands a slow, modest, and careful design process – a cautious Prometheus, as Latour puts it (see Latour 2008).

12 See Ingold 2011 for the concept of meshwork.

13 See *The Orkney News* 14 March 2017.

14 See Puig de la Bellacassa 2017 for the concept of living technology.

REFERENCES

Adam, B., and C. Groves, *Future Matters: Action, Knowledge, Ethics* (Leiden: Brill, 2007).

Anshelm, J., *Mellan frälsning och domedag: om kärnkraftens politiska idéhistoria i Sverige 1945–1999* (Eslöv: Symposion, 2000).

ARC, *Amager Resource Centre*, <https://www.a-r-c.dk/> [accessed 11 November 2019].

Bakhtin, M., *The Dialogic Imagination: Four Essays*, ed. by M. Holquist (Austin: University of Texas Press, 1981).

BIG Vortex, *A Building-Site Art Installation by realities: United for Amagerforbrænding, Copenhagen*, 12 April 2012, <https://www.youtube.com/watch?v=_GL3xAaIcvI> [accessed 11 November 2019].

Bladh, M., *El nära och långt borta: hur kan hushållen agera på elmarknaden?* (Linköping: Linköping University, 2007).

Brown, N., B. Rappert, and A. Webster, *Contested Futures: A Sociology of Prospective Techno-Science* (Farnham: Ashgate, 2000).

Bowker, G. C., 'Second Nature Once Removed. Time, space and representations', *Time & Society*, 4.1 (1995): 47–66.

Bredsdorff, M., 'Leder: Danmark som Europas skraldespand', *Ingeniøren*, 11 October 2013, <https://ing.dk/artikel/leder-danmark-europas-skraldespand-162456> [accessed 11 November 2019].

Bredsdorff, M., and S. Wittrup, 'Sådan blev Amager Bakke alligevel gigantisk', *Ingeniøren*, 4 September 2015a, <https://ing.dk/artikel/saadan-blev-amager-bakke-alligevel-gigantisk-178430> [accessed 11 November 2019]

— 'Trods advarsler: Københavnere får milliardregning for Amager Bakke', *Ingeniøren*, 4 September 2015b, <https://ing.dk/artikel/trods-advarsler-koebenhavnere-haenger-paa-milliardregning-amager-bakke-178433>[accessed 11 November 2019].

Brown, N., B. Rappert, B, and A. Webster, *Contested Futures: A Sociology of Prospective Techno-Science* (Aldershot: Ashgate, 2000).

Dagbladet Information, 'Hvad skal væk? Barsebäck', 30 May 2005, <https://www.information.dk/indland/2005/05/vaek-barseback> [accessed 12 November 2019].

DR (Danish National Radio and TV), 'Spild af dine penge', 18 April 2017, <https://www.dr.dk/tv/se/spild-af-dine-penge/spild-af-dine-penge-saeson-2/spild-af-dine-penge> [accessed 12 November 2019].

Edwards, P. N., 'Infrastructure and Modernity: Force, Time, and Social Organization in the History of Sociotechnical Systems', in T. Misa, P. Brey, and A. Feenberg, eds, *Modernity and Technology* (Cambridge, MA: MIT Press, 2003), pp. 185–225.

Ellers, H., 'Vølund hev storordre i land med lobbyarbejde', *Metal Supply*, 9 October 2012, <https://www.metal-supply.dk/article/view/89395/volund_hev_storordre_i_land_med_lobbyarbejde?ref=newsletter> [accessed 12 November 2019].

Energy of Orkney (2017) Orkney Sustainable Energy Strategy 2017/2025, <http://www.oref.co.uk/wp-content/uploads/2017/10/Orkney-Sustainable-Energy-Strategy-2017-2025.pdf>.

European Union, *Refuse Derived Fuel, Current Practice and Perspectives (B4-3040/2000/306517/MAR/E3)*, July 2003, <http://ec.europa.eu/environment/waste/studies/pdf/rdf.pdf> [accessed 12 November 2019].

Flaws, M., and G. Lamb, *The Orkney Dictionary*, 2nd ed (Kirkwall: The Orkney Language and Culture Group, 2001).

Ford, R., '"Lokkars! It's Thomas o Quoyness": The Role of Humour in the Dialogical Negotiation of Cultural Identity in Orkney' (Dissertation, University of the Highlands and Islands, 2013), <https://www.academia.edu/15570341/_Lokkars_Its_Thomas_o_Quoyness_The_Role_of_Humour_in_the_Dialogical_Negotiation_of_Cultural_Identity_in_Orkney> [accessed 12 November 2019].

Franklin-Wallis, O., 'Think Bigger: Bjarke Ingels on Why Architecture Should Be More Like Minecraft', *Wired*, 13 September 2016, <http://www.wired.co.uk/article/architect-bjarke-ingels> [accessed 12 November 2019].

Grundtvig, A., 'Skibet er ladet med skotsk affald – til forbrænding på Amager Bakke', *Politiken*, 4 November 2017, <https://politiken.dk/forbrugogliv/art6187640/Det-f%C3%B8rste-skib-med-udenlandsk-skrald-er-ankommet-til-Amager-Bakke> [accessed 12 November 2019].

Gubrium, J. F., and J. A. Holstein, *Analyzing Narrative Reality* (London: Sage, 2009).

Habermann, N., 'Amager Bakke starter nedtælling' *Magasinet Kbh*, 5 March 2013, <https://www.magasinetkbh.dk/indhold/amager-bakke-arc> [accessed 12 November 2019].

Haraway, D., 'Situated Knowledges: The Science Question in Feminism and the Privilege of Partial Perspective', *Feminist Studies*, 14.3 (1988): 575–599.

— *Staying with the Trouble: Making Kin in the Chthulucene* (Durham, NC: Duke University Press, 2016).

Hegelund, S., and P. Mose, *Lobbyistens lommebog – politikere under pres* (Copenhagen: Gyldendals Forlag, 2013).

Hughes, T. P., *Networks of Power: Electrification in Western Society, 1880–1930* (Baltimore: John Hopkins University Press, 1993).

HeHe, *Nuage Vert*, 2008, <http://hehe.org.free.fr/hehe/NV08/index.html> [accessed 12 November 2019].

Ingels, B., *Yes is More – An Archicomic on Architectural Evolution* (Köln: Evergreen, 2009).

'Worldcraft', *Future of Storytelling*, 2012, Video, <https://futureofstorytelling.org/video/bjarke-ingels-worldcraft> [accessed 12 November 2019].

Ingold, T., *Being Alive: Essays on Movement, Knowledge and Description* (London: Routledge, 2011).

Jasanoff, S., and S. Kim, 'Sociotechnical Imaginaries and National Energy Policies', *Science as Culture*, 22.2 (2013): 189–196.

Jensen, C. B., *Ontologies for Developing Things: Making Health Care Futures Through Technology* (Rotterdam: Sense Publishers, 2010).

Kall, A., 'Förnyelse med förhinder: den riksdagspolitiska debatten om omställningen av energisystemet 1980–2010' (PhD Thesis, Linköping University, 2011), <http://liu.diva-portal.org/smash/get/diva2:398531/FULLTEXT01.pdf> [accessed 12 November 2019].

Københavns Kommune, 'Kbh 2025 Klimaplanen – En grøn, smart og CO2-neutral by', 2012, <http://kk.sites.itera.dk/apps/kk_pub2/index.asp?mode=detalje&id=930> [accessed 12 November 2019].

Koolhaas, K., '100 Most Influential People – Bjarke Ingels', *TIME Magazine*, 21 April 2016, <http://time.com/collection-post/4301248/bjarke-ingels-2016-time-100/> [accessed 12 November 2019].

Laird, F. N., *Solar Energy, Technology Policy and Institutional Values* (Cambridge: Cambridge University Press, 2001).

Lange, M. A., *The Norwegian Scots: An Anthropological Interpretation of Viking-Scottish Identity in the Orkney Islands* (Lampeter, Wales: Edwin Mellen Press, 2007).

Larkin, B., 'The Politics and Poetics of Infrastructure', *Annual Review of Anthropology*, 42 (2013): 327–343.

Larsson, S., *Regera i koalition: den borgerliga trepartiregeringen 1976–1978 och kärnkraften* (Stockholm: Bonnier, 1986).

Latour, B., 'A Cautious Prometheus? A Few Steps Toward a Philosophy of Design (with Special Attention to Peter Sloterdijk)' [Lecture at *Networks of Design* conference, Design History Society, Cornwall, 3 September 2008], <http://www.bruno-latour.fr/sites/default/files/112-DESIGN-CORNWALL-GB.pdf> [accessed 12 November 2019].

Latour, B., and P. Weibel, eds, *Making Things Public: Atmospheres of Democracy* (Cambridge, MA: MIT Press, 2005).

Lindquist, P., 'Det klyvbara ämnet: diskursiva ordningar i svensk kärnkraftspolitik 1972–1980' (PhD Thesis, Lund University, 1997).

Lindström, K., and Å. Ståhl, *Patchworking Publics-in-the-Making* (PhD Thesis, Malmö University, 2014), <https://dspace.mah.se/bitstream/2043/16093/2/Patchwork_Lindstrom_Stahl_2014_korrigerad.pdf> [accessed 12 November 2019].

Marres, N., *Material Participation: Technology, the Environment and Everyday Publics* (London: Palgrave Macmillan, 2012).

Martini, J., and N. Sandøe, 'Ida Auken raser over direktionen på Amager Bakke: Det er uacceptabelt langt forbi det rimelige', *Finans*, 24 August 2016, <http://finans.dk/protected/finans/ECE8946386/ida-auken-raser-over-direktionen-paa-amager-bakke-det-er-uacceptabelt-langt-forbi-det-rimelige/?ctxref=ext> [accessed 12 November 2019].

Martinson, H., *Aniara: en revy om människan i tid och rum* (Stockholm: Bonnier, 1956).

Matarasso, F., 'Stories & Fables: Reflections on Cultural Development in Orkney' *Highlands and Islands Enterprise*, 2012, <http://www.hie.co.uk/regional-information/economic-reports-and-research/archive/reflections-on-cultural-developments-in-orkney.html> [accessed 12 November 2019].

Miljøministeriet, 'Danmark uden affald: Genanvend mere – forbrænd mindre', *Miljøministeriet*, 2013, <http://mst.dk/media/130620/danmark_uden_affald_ii_web-endelig.pdf> [accessed 12 November 2019].

Moser, I., 'Making Alzheimer's Disease Matter. Enacting, Interfering and Doing Politics of Nature', *Geoforum*, 39.1 (2008): 98–110.

Nye, D., *Electrifying America: Social Meanings of a New Technology, 1880–1940* (Cambridge MA: MIT Press, 1992).

Pedersen, M., 'B&W Vølund vinder af Amagerforbrændings milliardprojekt', *Energy Supply*, 20 January 2012, <https://www.energy-supply.dk/article/view/75334/bw_volund_vinder_af_amagerforbraendings_milliardprojekt> [accessed 12 November 2019].

Puig de la Bellacasa, M., *Matters of Care: Speculative Ethics in More Than Human Worlds* (Minneapolis: University of Minnesota Press, 2017).

Rohracher, H., 'The Role of Users in the Social Shaping of Environmental Technologies', *Innovation: The European Journal of Social Science Research*, 16.2 (2003): 177–192.

Schröder, K., *BIG TIME – Documentary about Bjarke Ingels* (Sunday Pictures, 2017).

Shakespeare, W., *The Tragedy of Hamlet, Prince of Denmark* (River Forest, IL: Aquitaine Media Corp., 1603/2009).

Shove, E., and G. Walker, 'Governing Transitions in the Sustainability of Everyday Life', *Research Policy*, 39.4 (2010): 471–476.

Stirling, A., 'Transforming Power: Social Science and the Politics of Energy Choices', *Energy Research & Social Science*, 1 (2014): 83–95.

Sylvast, A., H. Hänninen, H, and S. Hyysalo, 'Energy in Society: Energy Systems and Infrastructures in Society', *Science & Technology Studies*, 26.3 (2013): 3–13.

The Orkney News, 'Local Fuel Poverty Charity Warns of Hike in Electricity Costs', 14 March 2017, <https://theorkneynews.scot/2017/03/14/local-fuel-poverty-charity-warns-of-hike-in-electricity-costs/> [accessed 12 November 2019].

Tsing, A. L., *Friction: An Ethnography of Global Connection* (Princeton: Princeton University Press, 2005).

Watts, L., 'The Orkney Electron: An Ethnographic Story' [Paper at *People Places Stories*, Linnæus University, Kalmar, 28 September 2011] <http://www.sand14.com/wp-content/uploads/2012/04/watts_OrkneyElectron_Sept2011.pdf> [accessed 12 November 2019].

Wiklund, M., I det modernas landskap: historisk orientering och kritiska berättelser om det moderna Sverige mellan 1960 och 1990 (PhD Thesis, Lund University, 2006).

3

PROPOSITIONAL POLITICS

*Endre Dányi and Michaela Spencer, James Maguire, Hannah Knox,
Andrea Ballestero*

INTRODUCTION

A CENTRAL PROMISE OF DEMOCRATIC POLITICS HAS BEEN THAT ALL ISSUES, MATTERS OF concern, or problems relevant for a political community can be dealt with in a standardised way. If climate change can be conceptualised as an issue at all, it is an issue that radically challenges this promise. As varying groups, communities and nations around the globe attempt to respond, each in their own particular way, we suggest – based on the ethnographic examples assembled here – that their modes of engagement offer us some important insights. These insights shed light on how people are attempting to make their societies more liveable in tumultuous times – commonly, yet contentiously, referred to as the Anthropocene – and, more particularly, how their responses to the predicaments they face outline an emergent form of thinking and doing politics. This form, we claim, is *propositional*.

While the term 'propositional' may at first strike a philosophical chord, the propositions that emerge from our four cases are not bound to the logical precepts of analytic philosophy, but are very much embedded within, and emerge from, specific material arrangements. The lineage of the term that we adopt is one coincident with the work of Bruno Latour. In his essay 'A Well-Articulated Primatology', Latour (2000) deploys 'proposition' as a counter-metaphor to think through the practices of experimental scientific setups. One of the primary features of scientific work, Latour argues, is to make parts of the world visible through particular apparatuses. His intervention is to suggest a move away from the optical metaphor of the 'gaze' to one of propositions. The gaze metaphor, he argues, is what weds us to an understanding of scientific practice as detached or objective, leading to the suggestion that the varying perspectives, thoughts or opinions we have about the world are merely biases that distort or colour our relationship to the object of study. With the optical metaphor, the only reasonable outcome one can strive for is to get rid of all these filters and see 'things as they are'. The only good gaze is one that is interrupted by nothing.

Latour asks what would happen to the subjects and objects of experimental scientific setups if, rather than understanding sets of ideas or opinions about the world as biases that distort, we understood them as resources with which to make new connections within and between parts of the world. What would happen, he muses, if we took, for example, 'feminist antipathy towards passivity' as a resource for world-making rather than as a force to be erased? Inspired by Donna Haraway's work, Latour invokes feminist critiques of female passivity as a way to think about, and open up, the range of other actors that science has attempted to pacify. While experimental setups have a tendency to configure the world through a *success versus failure* mode of innovation, or through a *fact versus fiction* narrative of knowledge production, a propositional mode opens up the world – its objects and materials – to modes of description that take 'Others' into account. It is here that non-humans enter the stage as an important part of such propositions.

A well-articulated proposition is an 'occasion given to entities to enter into contact', or an 'interpretative offer' for non-humans to act in ways that might surprise us. Latour draws upon primatologist Thelma Rowell's work with sheep to emphasise the point.

> I tried to give my sheep the opportunity to behave like chimps, not that I believe that they would be like chimps, but because I am sure that if you take sheep for boring sheep by opposition to intelligent chimps, they would not have a chance. By placing them, quite deliberately and quite artificially, into the paradigm of intelligent chimps, I gave them a chance to express features of behaviour hitherto unknown. The more I work at it, the more autonomous my sheep may become. (Latour 2000: 372)

The proposition creates, in this instance, an opportunity for sheep to act otherwise, while at the same time leaving their participation open-ended. This is not humans experimenting on the world but collaborating with it and responding to it. What would happen if we articulated propositions that gave, for example, sheep the 'opportunity to behave like chimpanzees'? In this setup, the more work a scientist does in imagining alternative capacities for sheep, the more independent sheep become.

This has much in common with recent work in the social sciences on the place of the 'experiment' as a contemporary method of social and political engagement (Morita and Jensen 2015; Karvonen and Heur 2014; Wilkie 2017; Corsín Jiménez 2017). We would like to suggest that a move from experiment to proposition might be one way of characterising analytically what we see occurring empirically in our respective ethnographic field sites. As uncertainty over the future becomes more pronounced, new modes of thinking and relating to one another, and to the varying and contested worlds we are part of, seem ever more urgent. As the more traditional dichotomies of the philosophy of scientific practice become irrevocably unsettled, and as other, non-Western modes of knowledge-making begin to gain more purchase, we are slowly seeing the emergence of more varied 'arts of living on a damaged planet' (Tsing et al. 2017). In line with these shifts, this chapter itself *proposes* a move from experiment to proposition as a less certain, more ambiguous, albeit more open and tentatively imaginative, mode of engaging with and relating to human/ non-human encounters.

In this regard, we believe that thinking propositionally is a timely intervention Similar to Latour, who reflects upon the reconfiguration of the subjects and objects of experimental scientific setups through the metaphor of proposition, we would like to pose a similar question. What would happen to the subjects and objects of political arrangements if we took various approaches to politics as resources with which to make new, and varying, connections to the world? Rather than viewing the political practices of others – and here we define 'others' very broadly – as experiments that can succeed or fail in accordance with more traditional renderings of what count as political outputs, we want to *propose* that thinking about politics more propositionally opens up the world in ways that are both ambiguous and promising.

As people continuously come into contact with a host of distressing environmental scenarios, they are being propelled into doing politics in ways that fall outside the more myopic definitions of what democratic politics is. A necessity to innovate, or do things differently, permeates multiple scales of society as traditional political remedies creak and groan under the weight of their standardised responses. But doing things differently contains risks; proposals to engage the world 'otherwise' are not always well received. Doing propositional politics is not a matter of explicating a set of political principles or ideas. It does not involve pedagogy in the traditional sense but is devised to set in motion materials, affects and processes. Latour's move is to think experimental setups propositionally. Building on this, we want to propose a more experimental mode of doing politics that we refer to as *propositional*.

In the four stories that follow – from Australia (Endre Dányi and Michaela Spencer), Iceland (James Maguire), the UK (Hannah Knox) and Brazil (Andrea Ballestero) – we have shaped our narratives in the form of a *proposition-in-emergence*. The way in which we render these stories supports the volume's overall commitment to experiment. However, while the title of this collection pushes us to think and describe *Energy Worlds in Experiment*, we adopt the performative spirit of proposition in an attempt to open up the boundaries of these terms. While the middle stories by James and Hannah engage with energy experiments in more explicit terms (green energy production and energy monitoring devices), the first and last stories by Endre and Michaela, and Andrea, respectively, address energy somewhat more implicitly as they consider how water collaborations and water promises can re-energise political experiments in the Anthropocene. In moving from experiment to proposition, and energy to Anthropocene, our aim is not to work against the grain of the collection's theme, but to treat it as an 'occasion to enter into contact' (Latour 2000).

The stories assembled here build upon one another, each bringing into view a new place and a new set of concerns about politics and experiment. After each author's contribution, we provide a transition commentary in a collective voice, designed to act as a bridge that connects the stories and reflects upon the move from experiment to proposition. In the first story, Endre Dányi and Michaela Spencer interrogate the politics of water in Australia. While in the context of the Anthropocene it is often assumed that different kinds of knowledges ought to be brought together, our first story problematises the politics occurring in the meeting between scientific experiments and other knowledge forms. In the process, the authors question the limitation of 'experiment' as a way of engaging with a world undergoing dramatic environmental change. This first story acts as a hinge upon which our other stories pivot, as we go on to explore other, more propositional, ways of world-making.

In our second story, James Maguire gives an account of how experimenting with volcanic landscapes in Iceland is triggering anthropogenic earthquakes, and risky propositions. This story makes explicit how the ongoing politics of experiment in the Anthropocene triggers non-human agencies that play an important part in propositional politics. As such, the story acts as a transition, or threshold, between the notions of experiment and proposition.

In our third story, Hannah Knox introduces us to home energy devices in the UK, and the data traces they produce. Here the author treats data propositionally. Reflecting on the various gaps that data produce – what data do not say but nonetheless force us to think about – generates the need for a response to the complex relationship between material energies and the social imaginaries they seem to reveal. The story extends our understanding of the role non-humans play in propositional politics, raising interesting questions about the relationship between knowledge and politics.

In our final story, Andrea Ballestero turns our attention to promise-making in the form of water pacts in Ceará, Brazil. Analysed as experiments, it is possible to see such pacts as failures. However, focusing on the potential of the pacts' form – as an aggregation of promises – allows the author to re-conceptualise them propositionally. As propositions, such aggregates cut the politics of difference differently, engendering alternate collectivities beyond more classic tropes of regional or national belonging.

Now let us turn to Milingimbi Island, off the coast of Australia's Arnhem Land, where Endre and Michaela introduce us to a politics of collaborative experiments.

POWER AND WATER ON AN ABORIGINAL ISLAND

Endre Dányi and Michaela Spencer

In recent years water has become a pressing issue for the Yolngu people living on Milingimbi Island off Australia's northern coast. Here, questions of water management are increasingly entangled with questions of how, and in what ways, anthropocenic futures might be enacted. A central claim of the Anthropocene is that the world has changed. This assertion often comes with the understanding that we (humans) are now experiencing an unprecedented era in global history, where science is charged with the responsibility of studying a thoroughly socialised nature. When, in 2016, both authors visited a water management workshop on Milimgimbi Island, the claim that 'the world has changed' was also strongly asserted.[1] However, we found that what followed from this claim was not at all straightforward or predictable.

FIG. 3.1 Power and Water Corporation banner, Milingimbi Island, East Arnhem Land

Yolngu recall earlier times in Milingimbi when family groups used to move from place to place between their homeland areas. When fresh water supplies began to turn salty, it was time to pack up and move to another place. Since then, responsibility for managing water supplies has been handed over to a utility company, Power & Water Corporation (PAWC). Under its management, the ability to access both electricity and water has become a central concern for many of the Yolngu on the island, as well as a political game for politicians who have promised to manage the issue in various ways leading into an upcoming election.

Our attention was drawn to the issue of Milingimbi's water when we, two ethnographers, attended a workshop that brought together Yolngu rangers, Traditional Owners (TOs) and a number of visiting scientists. The purpose of the meeting was to enable visiting scientists (mostly German hydro-geologists) to engage with relevant Yolngu authorities and outline a collaborative research project in Milingimbi. The workshop started inside the office of the Crocodile Islands Rangers, an independent Yolngu ranger group working on land and sea management in Milingimbi and surrounding homeland areas. The participants then moved to a series of important locations dotted across the island – the water tower which held the community domestic water supply, a billabong located a little way out of town, and a proposed site for several aluminium measuring towers, each around five meters tall.

Only a few days before this workshop, there had been other visitors to the island, including a local Member of Parliament (MP). This MP was on the campaign trail, and had talked a lot about building new houses and housing extensions for local people who for several years had had new developments halted because of restrictions to water supplies. As the workshop in the rangers' office began, the conversation turned almost immediately to questions about housing. How much water was in the aquifer? Could new houses be built soon? How could the scientists help? Questions were thrown forward by Yolngu workshop participants, who hoped that the scientists could assist them with pressing issues around housing and overcrowding. However, the scientists were clear from the outset that their project was about research alone – water quality, salinity and transpiration rates – not about politics. They were on the island because climate change had radically altered the available water on Milingimbi. New – scientific – data were needed to understand the island's hydrological flows and assess water management problems before beginning to think about what to do next.

Following this statement there was some tension in the room, which remained unresolved as we packed our bags and drove out to various field sites. The first stop was a water tower – a large, elevated tank supplying the Milingimbi township. The scientists began talking to the rangers about the levels of water in the tower, the frequency of Milingimbi's rainfall and the transpiration rates of the surrounding vegetation. As this group of men – scientists and Yolngu rangers – stood in the hot sun, looking up at the tower, the two of us found some relief in the shade, sitting with a group of TO women, who were likewise marvelling at the willingness of the scientists to come here, stand in the sun and get things done the hard way. If they wanted to know about water, these women suggested, all they had to do was ask. They would have been happy to show them.

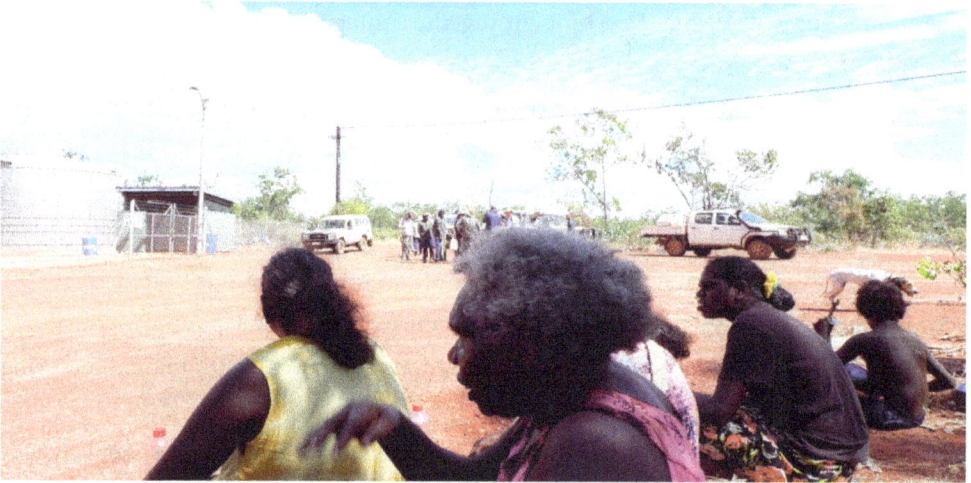

FIG. 3.2 Yolngu Traditional Owners and Rangers meet with scientists near the water tower, Milingimbi Island, East Arnhem Land

When we moved down the road to a billabong, the scientists were interested in the salinity of Milingimbi's surface water. There, as the scientists were getting to work, the Yolngu stood at the water's edge telling stories. To us, as interested bystanders, these differing sets of practices seemed to coexist quite easily, but they also revealed some tensions. While the head scientist threw in a salinity meter, one of the Yolngu present suggested that she could have taken a reading much more quickly – simply by dipping in a finger and tasting the water.

As the scientists talked about how the water was too salty for fish, another woman at the edge of the group pointed to the remnant ashes of a fire where her son had caught a barramundi fish, cooked and eaten it just a few days earlier.

By the end of the day we arrived at our final stop: a swampy area where many paperbark trees were growing. The discussion centred on one of the measuring towers that the scientists wanted to erect, and about how it might be kept safe from fires and local children.

FIG. 3.3 Meeting at Nilatjirriwa (billabong/waterhole), Milingimbi Island, East Arnhem Land

FIG. 3.4 Yolngu Traditional Owners and Rangers meet with scientists at the proposed site for one of the measuring towers, Milingimbi Island, East Arnhem Land

The towers offer a way of measuring transpiration from vegetation across the island. They are sensitive pieces of equipment that need careful monitoring. After significant rainfall, their collection buckets would have to be emptied to preserve the integrity of further measurements, and data would have to be collected and recorded continuously. The lead scientist had the components of a tower with him, and talked about installing it soon. At this point, however, conversation suddenly came to a halt. Permission for such installations was not something that the Yolngu ranger group, who the scientists had been speaking to all day, could give alone. It would also depend on the outcome of negotiations between TOs regarding the character and boundaries of their lands. While the scientists had been focused on the sampling practices and locations required for their transpiration measurements, they had failed to notice that the day's events were a series of meetings and negotiations where the local authorities were not the Yolngu rangers – almost exclusively men – but the TO women, who were careful to stay in the shade all day.

At the beginning of this meeting of scientists, Indigenous rangers and TOs, the assertion was made that the world had changed, and new data were needed to record these changes. To us, as visiting ethnographers, the invocation of climate change as a means to seemingly disregard or reset the knowledges of the Yolngu TOs was perplexing. By tracing the day's events through this empirical story, we do not seek to question the existence of climate change but suggest that what follows from it is far from straightforward. By abiding with the apolitical assumptions accompanying their climate change research, the scientists and the utility company missed much of what was at stake in the journey we took out 'on country', and the politics on display throughout the workshop. It is exactly this political work that our telling of this story seeks to bring to the fore, highlighting these negotiations in the face of claims being made under the banner of the Anthropocene, both here, and in places far removed from Milingimbi Island.

OTHERING EXPERIMENTATION

Peeking out from behind obvious modes of doing politics dependent on the production of 'objective scientific data', another can be seen lurking within these events in Milingimbi: a mode of doing politics that becomes visible as the Yolngu elder women sit quietly in the shade. As the sun beats down, and scientists and Yolngu rangers stand in a huddle, talking about water quality and the future of Milingimbi community, the women's comments on the unfolding events seem to point to other possible practices and forms of expertise.

Reading the events of the water workshop at Milingimbi through an STS lens attentive to difference in knowledge practices, certain political and epistemic practices come into focus (Blaser 2009; Law and Ruppert 2016; Stengers 2011; Verran 1998). Clearly, the hydro-geologists visiting the island brought with them an assumed set of arrangements for doing knowledge *and* politics. This was a familiar set of arrangements, which emerged with modern experimental science and which has, for a long time, enjoyed a priority position in democratic politics (Ezrahi 2012; Thorpe 2016). The hydro-geologists recognised there was considerable uncertainty associated with the quality and quantity of groundwater on Milingimbi Island. For this reason, they thought it was necessary to carry out very specific measurements, associated with a very specific regime of scientific testing. Testing the salinity of surface water (at the billabong) and the rate of transpiration of groundwater (through the island vegetation) were all part of this regime.

In establishing the bounds and practices of experimental science as a means to engage with the potential futures of Milingimbi water, however, the scientists encountered some problems. The ground, where they were working, was not simply part of one uniform globe, which implicitly locates and legitimates the conduct of modern Western science and demo-cratic decision-making. Here, ownership and the work of legitimating the foundations of knowledge was far more complicated, ambiguous and contested. Within the practices of the workshop, the claim that 'the world has changed' failed to reference, in any straightfor-ward manner, a given or accepted state of affairs. Rather, the claim emerged as a contested proposition – an attempt to produce a new baseline legitimating the maintenance of old sets of modern knowledge-making practices and reinforcing a version of democratic politics where decisions always happen elsewhere (in the city, a committee meeting, in Parliament, in conversations that render the difficulties discussed above invisible).

By drawing attention to this fraught and problematic move, which asserts an anthro-pocenic reality as the new grounding for scientific inquiry and future-making on a global scale, this opening ethnographic story sets the scene for an inquiry into a different – propo-sitional – mode of politics, which is to follow.

Asserting 'the world has changed' as a new condition initially seemed to ground and legitimate further rounds of experimental enquiry in a modern and familiar mode. However, it was this move that the women by the water tower quietly resisted, as they subtly hinted at a broader political forum that existed in and around the water tower discussions. That forum became fully fledged at the final site visit of the day, as the TOs resisted the unproblematic imposition of a measuring tower and scientific apparatus without proper negotiation over the ownership of the land and the responsibilities for remuneration and care that it entailed.

By opening with this story, our suggestion is that the invocation and inhabitation of Anthropocene worlds is beginning to provoke not only the maintenance of particular epistemic problems and practices, but also new forms of epistemo-political action and resistance – forms of politics that show up and slow down the conduct of business as usual (de la Cadena 2010). This is a politics that calls out the blind spots of existing mainstream epistemic practices and offers opportunities for more generous and appropriate imbroglios of action, which hold in tension questions of *who* knows, *what* is known and *how* we know it, so as to enable different answers to questions about where, how and through what diverse sets of relations we might continue to live. To explore these issues further, the next story turns to a rather different kind of encounter – this time between geothermal energy and the town of Hveragerði in Iceland. This shift in geographical and empirical focus helps to deepen our appreciation of the variety of ways in which such imbroglios might become manifest, and the kinds of issues and concerns that might surface.

SHAKY MATTERS: POLITICS AT THE THRESHOLD

James Maguire

Over the course of its history, Iceland has been plagued with various forms of instability: topographic, climatic, and more recently, financial. In an effort to move beyond these vicissitudes, the Icelandic state has turned towards the landscape, converting the topographic instability of some of the country's vast and powerful volcanic zones into economically productive sites of geothermal energy. This is done mainly for the provision of electricity for the aluminium industry. Proponents argue that coupling one of modernity's primary metals with one of the world's greenest energy forms is a sign of planetary, as well as industrial, progress. However, the conversion of volcanic forces into energy resources is having some disturbing effects.

Making geothermal energy is a risky affair. As drills penetrate deep underground in search of volcanically heated fluids hot enough to drive electricity turbines, they have begun to produce troubling collateral effects. Anthropogenic earthquakes have emerged as a feature of life for the residents of Hveragerði, a small town on the outskirts of the volcanic area. A friend and town resident characterised the situation to me one day as a 'shaky matter', a term that could refer both to the physical disturbances the town has to endure as the earth continues to tremble, and also the instability of the ethical and political assumptions that underly the project.

FIG. 3.5 View of the Hellisheiði Geothermal Power Plant within the Hengill Volcanic Landscape.

This 'shaky' matter has generated its own unique political arrangement, as the residents of Hveragerði try to come to terms with the municipal energy company producing these anthropogenic earthquakes. Part of this arrangement was the implementation of a predictive seismic warning system. This system – designed and operated by the Icelandic Meteorological Office – was intended to predict the possibility of increased seismic activity in the area, as well as notify the town council in the event of potentially dangerous levels.

The town's mayor suggested to me that warnings are a residual issue of a larger political discussion going back to the establishment of the geothermal power plant. While Hveragerði is the closest residential area to the power plant, the town is not the lease holder of the land on which the power plant operates. As a result, it has almost no formal political remedies at its disposal. Warnings, it seems, are the best it can get in shaky circumstances.

One of the first warnings to be issued became a highly public affair: the prediction of a possible 5.2mw earthquake within a seventy-two-hour timeframe. While only very minor tremors were felt over the course of the next week, the negative media attention generated by the warning, not to mention the uncertainty and anxiety it caused the residents, prompted the town to rethink its approach.

Today, such warnings consist of a simple electronic communication between the power plant and Hveragerði's town council. While these warnings are subsequently posted onto the town's website, residents and media outlets have only a vague awareness of their existence. But what type of political arrangement is this, in a situation of ongoing earthquake production? Town council members and residents talk of being in 'an awkward situation', and of 'needing to dance on a line' while 'not shouting too loud or attracting too much attention to the warnings'. On the one hand, there is the political necessity to publicise these warnings – public safety – while on the other, there is the political fear that drawing too much attention to them could also be damaging for the town.

There are several reasons for this, but the main one is the shadow cast over the town by the 2008 financial crisis. For many years Hveragerði used the bountiful supply of steam and hot water emanating from the fractured earth to develop a large horticulture industry, and was known for many decades as the 'greenhouse capital of Iceland'. However, the property crash that followed the crisis brought this industry to a grinding halt. In the wake of the crash, the town survives from a few main businesses, two of which are a care home for the elderly and a rehabilitation centre for people with chronic illness. Publicising seismic warnings is highly problematic for a town with such health-based industries.

While earthquake warnings emerge from geological instability, they also have the potential to generate other forms of instability, which although beyond the seismic are nonetheless bound up with it. Making too much of an issue out of these warnings can bring about effects equally, if not more, destabilising than the seismic instabilities surrounding them. Dwindling investment and property prices, as well as increased unemployment, are realities that such a town cannot sustain.

My geologist friends at the municipal energy company describe anthropogenic earthquakes through the language of thresholds: geothermal activity is accelerating longstanding seismic rhythms, triggering earthquakes through the production of new thresholds. Etymologically, 'threshold' comes through Greek, meaning 'at the door'. Being a portal through which we move between places, a door is a helpful way to think of a threshold as a transition. Here the many metaphors that invoke the crossing of a threshold as a symbolic transition come to mind. But the thresholds being triggered in these volcanic landscapes are more than crossings though symbolic space: they are akin to phase shifts, moments of critical and sudden change through which new states emerge.

Much as geothermal activity triggers seismic thresholds, warnings trigger political ones; these are moments when warnings change state and become potentially more destabilising than that which they warn against. Maybe we can think of these warnings propositionally. As mediated earth signals they are a way of understanding the shaky earth as a risk proposition. They are in some sense 'interpretative offers', made through human-earth arrangements to relate to the geological, politically (Latour 2000: 15). But what they *propose* is ambiguous; their language is one of probabilities and likelihoods, with no clearly delineated course of action. In this sense, they remain open, even underdetermined. Relating to this proposition – either *making an issue* out of anthropogenic earthquakes (publicise the warnings widely and suffer the likely economic consequences) or *not making an issue* out of them (render them next to invisible on the website and take the consequences of inaction) – is a tricky political dilemma. In fact, it opens up more explicitly the paradox of *not making an issue out of something* as a legitimate political response. Maybe this is one of the conundrums of living on, and with, a damaged planet, as options for change increasingly come with their own set of unpalatable dilemmas.

THE ENERGETIC EARTH AS CONSTITUTIVE OTHER

James's story, akin to Endre and Michaela's, also engages the relationship between experiments and politics. In this instance, we get a glimpse of the ambiguous politics at stake when green energy dreams are materialised through infrastructural experiments in volcanic landscapes. While our first story questions the limitation of *experiment* as a way of engaging with the world under conditions of environmental distress, this story makes more explicit how experiments in volcanic landscapes trigger risky *propositions*.

But such propositions are not an entirely human affair. The move made here tentatively points towards *the earth* as also having a say in how propositions emerge and take form. While Latour suggests that the ways in which scientists configure their propositions – either as 'well' or 'badly' articulated – is what affords nonhumans a more, or less, extensive role in how they come to participate in our world, this piece offers a slightly different reading. Here, nonhuman participation arises through human interventions, while also shaping the configuration of human modes of political engagement (Latour 2000; Lovelock 2000; Latour 2017). In this rendering, the articulations of non-humans can also impact upon the configuration of human modes of political engagement. As seismic thresholds are triggered, warnings, as mediators, trigger political thresholds that put this small town in south Iceland in a difficult bind. As unsavoury as this bind is, the town has to decide how to contend with the propositions emerging from geo-political arrangements.

Highlighted here is the nature of the political in a world increasingly striving for greener energy infrastructures. In some ways, this story serves as a counter-narrative, a rejoinder perhaps, to the implicit complacency that can arise with so-called green energy innovations. While from the perspective of carbon metrics, geothermal energy is one of the greenest forms of energy on the planet, the extractive and financial logics at play in south Iceland are not dissimilar to those of the fossil fuel industry. Maybe, then, it should come as no surprise that the burden related to renewable energy transitions is delegated to those with limited power or influence. The question of who bears the political, economic and social brunt of the risks that come with energy transitions is one that clearly needs to be addressed. We see this with the politics of fracking in the US, as well as energy extraction for resource use around the globe. In this specific case, the small town finds itself located at the conjuncture of a municipal energy company (Reykjavik Energy) and a large multinational aluminium company (Century Aluminium). While both sides argue that coupling one of modernity's primary metals with one of the world's cleanest energy forms is a sign of progress, such progress looks slightly distorted from the situated perspective of the town.

While warnings can show a shaky earth to be a proposition about risk, they are also a risky proposition that the town has to somehow respond to. Relating to this proposition – as the text outlines – opens up more explicitly the paradox of *not making an issue out of something* as a legitimate political response. Can we call this, *contra* Marres (2005), non-issue politics or a form of anti-politics? At a minimum, such a proposition puts the very question of politics at stake, as it opens the question of how much politics any given matter can bear.

While Jacques Rancière suggests that the construction of the domains of the political and the non-political is, in essence, the definition of politics, the above form of propositional politics works to remind us of the complexity of our political configurations in the Anthropocene (Rancière, Bowlby, and Panagia 2001). In south Iceland, such propositions, like the energetic matter they emerge from, are shaky. They set in motion a series of dilemmas that are hard to contain within a more traditional definition of politics. As the propensities and forces of the earth are mobilised for greener futures, we might stop for a moment to ask how, as we re-energise our infrastructures, we might also re-energise our politics in a manner that offers ways out of risky propositions.

CLIMATE CHANGE AND THE POLITICISATION OF THE MUNDANE

Hannah Knox

Several of the political challenges of climate change derive from its identification as a global problem with anthropogenic causes. Climate change is generally framed as the problem of the impact that human beings, as a species, are having on the world climate system. Rendering the problem in this way implies that everyone in the world is equally responsible for, and equally affected by, climate change. Many have critiqued this view, for instance Donna Haraway and Dipesh Chakrabarty, but my aim here is not to dwell on whether it is right or wrong but to consider what we might call a material-semiotic rendering of the problem of climate change as an 'ethnographic fact', in order to think about what this manifestation of anthropogenic climate change does for the possibilities of political action (Haraway 2016; Chakrabarty 2017).

As I have explored elsewhere, one consequence of identifying 'human beings' as the cause of climate change is that the question 'What is a human effect?' becomes open to scrutiny (Knox 2014). Far from singularising our understanding of humans as a species, I suggest that anthropogenic climate change prompts a cutting and splicing of the generic human into a social geography of proportional responsibility. Here the agents of environmental degradation proliferate across orders of being and scales of action – manifesting variously as the public, the individual, the food system, energy networks, oil corporations or the president of the United States.

One central node around which this social geography of proportional responsibility revolves is the energy infrastructures that have led to global climate change. It is all well and good to call for an 'energy transition' from fossil fuels to renewable energy sources, but those involved in activities to bring this about are clear that this requires more than utopian visions of a clean, green future. Even to know how to move towards this future requires work to understand and map the actually-existing relations that constitute the energy system within which people are entwined. Let us turn to an ethnographic account of one project that has been trying to use energy data to effect this kind of political intervention.

The home energy monitor sits silently in the corner of the living room. There is a certain pride in having constructed something like it. We crafted it ourselves, at an energy workshop – a group of Manchester residents learning together about environmental monitoring, cooperative action, climate change and energy infrastructure. We started out with a small clear plastic bag of components: a circuit board, wires, LEDs, a soldering iron, a computer and a screen. People who had taught themselves, through trial and error, how to make an energy monitor told us how to solder the components to the circuit board to enable pulses of electricity to flow and be detected. We equipped our units with electrical pulse sensors, also soldering them in place. We fitted temperature sensors. We didn't really know how these sensors worked. All we knew was that they came from manufacturers in China and that it was easier to source electricity sensors than sensors for gas.

FIG. 3.6 Equipment for making an energy monitor

It was hard to make the energy monitor work. Sometimes the connections that we soldered were not good. It takes skill to solder a board. But with some help from one another, sharing soldering irons, trying different kinds of solder, most of us got there. After eight hours of melting, aligning, sticking and programming, I took my monitor and receiver home and clamped them to the wire that goes into my electricity meter. There were two wires. I didn't know if it mattered which wire I clamped them to. I turned on the monitor, but it remained blank. I followed the instructions on how to set up an online portal to see my energy flow – and there it was. Numbers appeared on the screen and I could see how much electricity was there. The energetic flows travelling along those wires were being translated into numbers. I never did get the actual screen of the energy monitor to work, but I could see my data online, and that meant I could begin to delve into what it meant.

In the next workshop, we began to look at our numbers. Stories proliferated. How were we to make sense of these data? Numbers were opening up cascades of relations. If we added the tariff that we were being charged (found on our energy bills), we could see how much money we were spending. But why did our energy cost this much? We could also calculate our carbon emissions. But how did the pulses of electricity moving through our meter link to carbon dioxide in the atmosphere? We worked out that it partly depended on the energy mix. Many of us said that we had green energy suppliers, and wondered how this affected the carbon emissions of the energy we were using. But it turned out that the electricity pulsing through our wires was just the same electricity as that used by our less ecologically minded neighbours. Being a green energy company customer, we realised, was more about invest-ment in the future of energy, rather than the consumption of a materially different product.

In this energy experiment, data provided a way of opening up the complexity of energy infrastructure. The energy data made by our home-built energy monitors was not a fixed point, a direct representation of an already-known entity. It was rather, in Stengers' (2014) terms, something that 'forced thought' and elicited a reaction. Data was good to think with. Not because it explained, but precisely because it often did not explain. Data here operated as a proposition, generating the need for a response to the complex relationship between material energies and social imaginaries that it seemed to reveal.

The relations highlighted by the data provided by the home energy monitor were, in one respect, relations between materials. But they were also relations between materials and people. Once one becomes aware of a house's energetic properties then one also realises that any change to the materiality of one's home (insulation, triple-glazing) inevitably has unforeseen consequences, and that action, intervention and behaviour come to matter anew. Inadequate energy conservation work leads to condensation – hence the mantra among housing retrofit experts: 'no insulation without ventilation'. But this condensation is not just the result of the house but also the people living in it. Even when energy-saving measures are installed properly, the conditions within which they exist are vulnerable to change. People hang washing on radiators and mould-spores grow on the walls.

Both numerical data and personal experience thus combine to make evident the rela-tionships between these phenomena – dry washing, damp walls. Ethnographic methods teach us to follow relations across such divisions, but so do these propositional forms of data science, which produce new forms of description and with them the promise of a new politics. Washing, mould, insulation, ventilation, become a matter not just of dirt, discomfort or health, but of responsibility, morality and infrastructure. Treated ethnographically, data traces about energetic relations allow us to tell different stories about the world that we and others inhabit. Data traces – whether numerical or ethnographic – open up energy. They help us 'learn more' about the social geographies of proportional responsibility. If this is a form of propositional politics, it is propositional not because it proposes what the future should look like, but because it emerges out of the relationship between the articulation of what is desired and the destabilising potential of what might become apparent in the tracing of relations.

TRUTHS AND POLITICS

Hannah's vignette opens up our exploration of propositions in two key directions. The first is the way in which it addresses the crucial role that particular relations – which we might call nonhuman – play in propositional politics. The second is the way in which it raises questions about how politics is able to proceed when it is not based on agreed truths but on an attention to the stories that can be told about social formations.

As James's story began to explore and as Hannah's elaborates, attention to physical matter is central to propositional politics. It is perhaps no coincidence that the propositional mode of 'relating to' comes into play exactly at that moment when understandings of the relationship between resources, natural processes and human needs are undergoing profound change. Latour, who also uses the language of proposition to describe possibilities for rearticulating the relationship between science and science studies, asserts that we need to better understand the crucial role played by nonhumans in world-making exercises. The question that follows is how do these non-humans 'contribute to how we imagine stories that no one without some level of intimate familiarity could make up?' (Latour 2000). For Latour, they contribute by participating in the creation of a proposition – a mode of description that necessarily emerges out of an intimate understanding of that which lies beyond ourselves. If in Latour's paper that which lies beyond is the world of primates and hyena sex organs, in Hannah's case it is a system of energetic relations which we can see enacted and represented through the building of an energy monitor and data visualisations. But while energetic relations, in Hannah's case, are enacted and represented, the agency of such nonhumans is much harder to grasp than anything we might relate to through the concept of 'the nonhuman actor'.

This propositional way of relating to materials, objects, systems and representations raises the question of how our modes of knowing are transformed, and what is done to the relationship between politics and knowledge. Clearly, in this formulation, doing politics is not a matter of explicating a set of political principles or ideas. It resembles pedagogy, but not in the traditional sense. It is devised to set in motion materials, affects and processes in ways that certainly involve learning but cannot be exactly understood as a form of didactic teaching. What we see in this story is the propensity of energy objects or techniques to open up and reveal certain things (and of course to obscure others). What is interesting is what these practices are expected to uncover. Indeed, what we find being addressed is not a field upon which predefined political agents might act ('the state must close the polluting power stations', 'private corporations must be held accountable'), but rather a re-description of a technical terrain (the properties of building materials, chemical signals, regulatory frameworks) in newly politicised terms.

There has been much talk recently of how we might be living in a 'post-truth' world, but the kind of political practice described here illustrates the ongoing centrality of evidence for what has been termed elsewhere as 'post-political' action. The difference is that the relationship between evidence and what is being evidenced now appears underdetermined. Inscription devices and infrastructures of knowing seem to have morphed into something quite other. The capacity of methods and 'others' (devices) to produce alternative forms of evidence as a way of reformulating political practice is the focus of our next story.

PROMISE AS EXPERIMENT: THE CUMULATIVE ENERGY OF AGGREGATION

Andrea Ballestero

Between 2008 and 2013 more than ten thousand people participated, one way or another, in the making of Ceará's Water Pact.[2] Ceará is a state in Northeast Region, Brazil, with severe water scarcity issues. The Pact was imagined as an experiment, a way of bringing society together to engage with water as a transversal material and ethical substance that requires a thorough commitment to care. The Pact would create this engagement by seeking to involve the greatest number of citizens and inviting them to make a public promise to care for water, asking them to participate in an open ritual of committing to the future of water from their particular context. The promises people made were registered on coloured slips of paper, briefly exhibited on the walls of community halls, schools, universities and churches. Once collected, they were converted into all sorts of electronic documents that circulated throughout the state as evidence of people's commitment to care. Ultimately, they were gathered into a series of Pacts at municipal, watershed and state levels.

FIG. 3.7 Ceará's semi-arid landscape during dry season

The organisers of the Pact hoped to interrupt history, to disrupt the usual ways in which the state addressed water issues. Those familiar responses included constructing new concrete infrastructures, creating and allocating new forms of water rights, and charging for bulk water. Rebecca, one of the Pact organisers, and a seasoned professional involved in land reform and participatory water projects for more than two decades, saw the historical conjuncture the Pact responded to in this way:

> We did it all, we applied the pacotinho [little package] of policies [e.g., decentralisation, economic valuation, participatory allocation of water] just as the international establishment recommends. Our State is a textbook example and, what happened? Fifteen years later, we have more carros pipa (water trucks) than before. So we needed to do something different. And this is where the Pact emerged. What is new these days in Ceará is the Water Pact.

Rebecca and her colleagues' hopes of doing things differently aimed to gather people, places and substances that could not be merged into a singular entity. They wanted to create a collective that not only tolerated, but asserted, the specificities and differences intrinsic to water issues throughout the state. They wanted to produce an intervention that kept alive, rather than flattened, the difference made by people's contextual circumstances. Rather than promoting belonging, the Pact was designed following a logic of aggregation. It was an open and never-ending accumulation, designed to avoid any alchemy that would transform the hundreds of promises into a single entity. The Pact was, and is, nothing more than the aggregation of promises. Its purpose continues to be to break with well-established political genealogies, to produce something that familiar political tactics could not have anticipated.

For many in Ceará, the Pact was a failure. Despite its magnitude – more than eight thousand participants made pledges to each other and to water – its results were not available for empirical verification. There was no Panopticon to oversee the enactment of the promises Pact participants made, no God's eye view from which to document the results. It was a composite whose contours and boundaries could not be precisely defined. Paradoxically, the Pact was an effort to produce difference, without having any way of verifying whether distinctions were systematically produced. This open-ended experimental character was its promise and its flaw – the Pact's source of hope and its historical fault.

FIG. 3.8 Participant documenting the Ceará's Water Pact negotiations for colleagues at his municipality

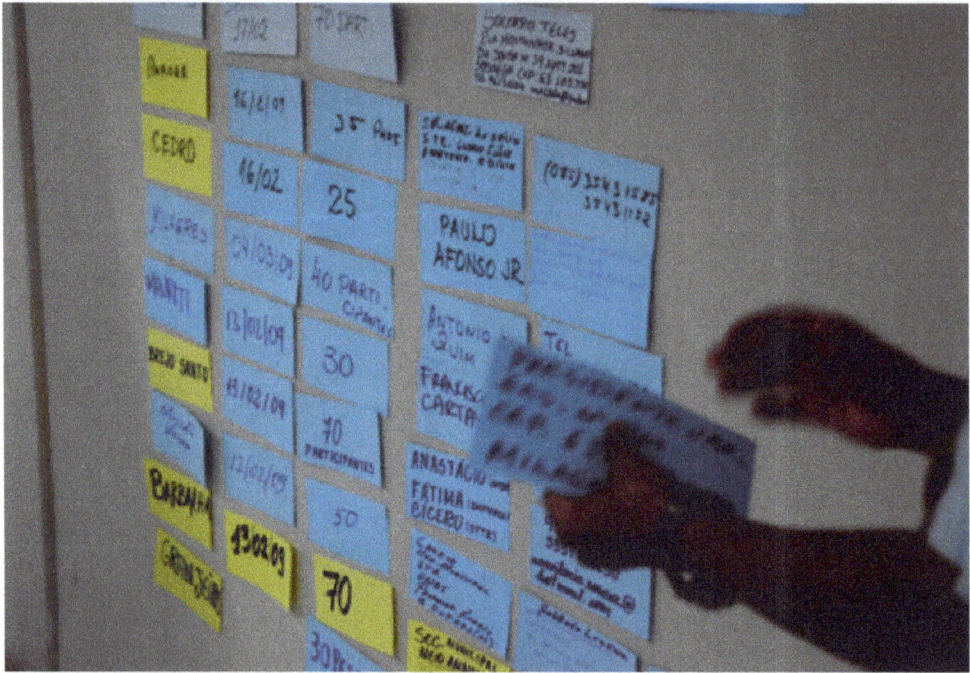

FIG. 3.9 Organising municipal Water Pact meetings where local promises would be made

In this ambitious effort, one humble participant carried an inordinate amount of weight. That nonhuman participant was a required piece, a necessary element for any aggregation to be accomplished. I found this participant at the core of the method that the consultants who led the Pact used. That participant is the coloured piece of paper. Yellow, blue, green and pink, these slips played different roles at different times. As people held them in their hands, quietly pondering what they could offer to the collective, the slips became the material embodiment of future care, the evidence of a form of collective energy. They were also evidence of past events, piled up as proof that this process had involved more citizens than any other dealing with water politics. They were exemplars of what a promise is, and also pedagogical aids necessary to disseminate the idea of a Pact and secure the 'right' kind of response from participants. The slips were records of promises to be, documents of a future that had not yet arrived.

But besides those multiple lives, all slips shared two characteristics. Firstly, each slip embodied one promise only. If any participant wrote more than one promise on a slip, it was returned and the author was asked to split it into two. Secondly, their material form, their dimensions and portability, allowed the coloured slips to be placed and displaced in different columns, on different walls, under different headings, in different stacks, at different community centres, on different desks. These coloured slips of paper were open to interconnections and tentative allocations; they were subject to correction during afternoon workshops or later reassigned to new scales as the Pact organisers 'systematically processed' the results of the four years of work they had invested in the Pact.

If 'practices and concepts come packaged together', what were the concepts packaged in the coloured slips of paper that hung from innumerable walls across dozens of locations in Ceará? (Rheinberger 1998) The promise/slip of paper brings practices and concepts together; it enables aggregation, a social form whose quantitative nature demands a qualitative reconsideration.[3] As a gathering of moral commitments, the Pact does not claim any form of unity. It is a plural construction, without any aspiration to singularity (Burge 1977; Link 2002). Merging promises into a tight, cohesive entity is intrinsically impossible. The aggregate is a practice, a concept, that adds without claiming to achieve unity. It selectively enrols something that is shared by all of its participants: the willingness to make a promise to care for water, without claiming to incorporate selves, subjects or citizens. The coloured slip of paper makes all of this possible.

With the Pact, I want to think about aggregates as a political proposition that can handle the proliferation of difference. The aggregate can be a social form that does not demand belonging. It sidesteps the problem of contradictory allegiances by not limiting the collective formations that people can be part of, but, on the contrary, allowing more loose and temporary groupings to be made and unmade according to different political, affective or epistemic affinities. For example, through endless databases built on dimensions (traits) of our social experience, we have become always already aggregated entities, regardless of our intention.[4]

How can we re-examine the aggregate as a political form open to the ethics of promise-making? What kinds of experiments might become possible if, rather than attempting to produce new holisms, we recuperated the world-making possibilities of aggregations (in their material specificities)? What kind of political energy might this unruly form yield?

The Water Pact had a peculiar affective charge, one that is shared by many of the political experiments being put together around the world to address complex and difficult issues such as climate change, energy transitions and securing universal access to water. That affective charge is at once experimental and melancholic. It entails an attempt to innovate, to propose a new way of dealing with things that are grounded in disappointment with previous modernist interventions and their claims around causality and certitude.

In the Pact, the slip of paper was a new experimental participant, something like a home-made Post-It. That piece of paper carried the weight of the promises each participant made. It standardised them very much in the way that liberal imaginaries of citizenship standardise different bodies. But what was standardised in the Pact was not a citizen; it was the pledge to do something in the future, to react to the legacy of the past. The slip of paper universalised that commitment to act, allowing a process of selective aggregation. The slip of paper made plucking a dimension of human experience, a promise, possible. This plucking procedure generated a flock of pledges, a multiplicity irreducible to a single unit because the promises were never intended to become cohesive.

Unlike more conventional liberal political forms, the Pact was a grouping built upon units connected in precarious ways. In this sense, the Pact shares more with the imaginary of digital relations and mobilisations than with forms of belonging to a kin group, community or nation. Despite its seemingly artisanal computational methods (it was all put together using Excel sheets and Word tables), the Pact operates following a digital imaginary of aggregation. It follows a propositional tactic that recuperates the value of what scholars tend to consider precarious social relations – ties that do not depend on a deep sense of embodied interpersonal responsibility. Maybe the Pact suggests that rather than pursuing robust social ties, strong connections and deep commitments, we need to reimagine what is possible when we mobilise fragile pledges, precarious ties and tenuous connections. Maybe the answers to many of the anthropocenic challenges we face lie less in heartfelt commitments and more in the potential of light promises.

In Ceará, but also elsewhere, people grow melancholy as they process their disappointment with previous modernist interventions and their causal claims. People long for the certainty of 'solving' problems that modernism promised. In the absence of that certitude, people seem open to departing from well-trodden roads into new territories, and to engaging in novel experiments. In this contradictory double affective state, simultaneously melancholic and experimental, people address what Dimitris Dalakoglou has called 'infrastructural gaps'; that is, the failures of the material interface between the state and the private sector (Dalakoglou and Kallianos 2014).

These gaps are clearly material. They record the inability of pipes, bridges and grids to deliver as they used to, or at least as it was promised they would. But such infrastructural gaps are also, and most profoundly, about the failure of the political imagination. They reveal the failure of the 'social contract' to organise matter, affect and will in the Anthropocene. The Water Pact emerges as a proposition within that gap. It is an experiment that requires huge amounts of money, time, energy and effort to be organised and enacted, yet it is not conceived as 'the most appropriate solution'. Its relation to certitude is radically different.

The Pact grows out of the need 'to do something different' without guaranteeing results with any precision. This is where the politics of experiment and openness show their potentially detrimental side. By renouncing any causal claims or certitudes, the Pact avoids fixing the ontology of water in any particular way – as a commodity, a spiritual substance, part of nature, or as a legal object. This ontological fluidity makes it difficult to allocate responsibility in any conventional way. For instance, given the absence of any central assessment or verification mechanism, the Pact challenges our expectations of causality and our desire to identify who is responsible for which consequence – who has responsibly fulfilled their promise and who has not. As a political proposition, the Pact does not definitively apportion responsibility to the state, the private sector or the individual. In fact, we could say it does the opposite. The Pact seems to disseminate responsibility vaguely among all sorts of familiar and new political actors. Ultimately, perhaps this openness makes the Pact a proposition that reflects the depoliticising effects of dominant political and economic schemes of the second half of the twentieth century (i.e. schemes we could call neoliberal).

But isn't openness and dissemination of responsibility something we desire when we attempt to reinvent what counts as political action? Aren't these open propositions particular 'arrangements of existence' that bring about their own possibilities for deranging and rearranging our worlds? (Povinelli 2016). The promises at the core of the Pact can be literally arranged and deranged. They can be moved across scales, pulled out, brought in, repeated. They can also be ignored or set aside, since there is no centralised body that could impose penalties for breaking them. This potential for radical reinvention and for inaction, for politicising and for depoliticising, seems inherent to propositional politics. Thinking the world otherwise, and venturing to rearrange our worlds, is never a clean task. It always risks mobilising unsavoury consequences. Politicising and de-politicising are not completely distinguishable from each other.

CONCLUSION: AGGREGATING A PROPOSITION-IN-THE-MAKING

All four pieces in this palimpsest of stories speak to the propensities of 'others' to re-energise our politics, the formation of new collectivities and the possibility of modes of description that relate the material and the imaginative. But these propositions, being open enough to treat the world as otherwise to the way we know it, are also risky; they hold promise, but are flawed, as the potential for failure resides in how they are arranged. Inaction, even irrelevance, are clear risks associated with doing things 'otherwise' in a political configuration that continues to motor along the highways of modernism. If propositional politics allows for failings and ambiguities, then it might be just the political form equivalent to the needed hypo-subjects of a new era.[5]

At this point, we can suggest something in a propositional spirit. We might entertain the possibility that rather than search for rearrangements that generate ontological clarity and unambiguous responsibility, we need more awareness of the inherently contradictory consequences of our political projects. Maybe we need propositions that are more self-conscious of their deranging potential – not as modernist risks to be avoided, but as constitutive turbulences that need to be embraced. Maybe it is worth staying in the shade, as Aboriginal women have taught us. Maybe we should commit to weak ties rather than searching for deep identifications.

NOTES

1 For a detailed account see Spencer et al. 2019.

2 For an extended analysis of this case see Ballestero 2017.

3 For two different ways of engaging with quantitative forms as qualitatively rich entities see Verran 2001 and Ballestero 2015.

4 For a discussion on the aggregate logic of the database as a governmental technology see Ruppert 2012: 126.

5 The hypo-subject – as opposed to the hyper-subject of the modern era – is a term that attempts to describe the human subject of a new climate generation; less masculinist and arrogant, more modest and humble. See Boyer and Morton 2016.

REFERENCES

Ballestero, A., 'The Ethics of a Formula: Calculating a Financial-Humanitarian Price for Water', *American Ethnologist*, 42.1 (2015): 262–278.

— 'Capacity as Aggregation: Promises, Water and a Form of Collective Care in Northeast Brazil', *The Cambridge Journal of Anthropology*, 35.1 (2017): 31–48.

Boyer, D., and T. Morton, 'Hyposubjects. Lexicon for an Anthropocene Yet Unseen', in C. Howe, and A. Pandian, eds, *Theorizing the Contemporary*, 1 January 2016 <https://culanth.org/fieldsights/hyposubjects> [accessed 7 November 2019].

— 'Hyposubjects.' Theorizing the Contemporary, *Fieldsights*, 21 January 2016 <https://culanth.org/fieldsights/hyposubjects>.

Blaser, M., 'The Threat of the Yrmo: The Political Ontology of a Sustainable Hunting Program', *American Anthropologist*, 111.1 (2009): 10–20.

Burge, T., 'A Theory of Aggregates', *Nous*, 11 (1997): 97–117.

Chakrabarty, D., 'The Politics of Climate Change is More Than the Politics of Capitalism', *Theory, Culture & Society*, 34.2–3 (2017): 25–37.

Corsín Jiménez, A., 'The Prototype: More Than Many and Less Than One', in A. Corsín Jiménez, ed., *Prototyping Cultures, Art, Science, and Politics in Beta* (London: Routledge, 2017), pp. 1–18.

Dalakoglou, D., and Y. Kallianos, 'Infrastructural Flows, Interruptions and Stasis in Athens of the Crisis', *City*, 18.4–5 (2014): 526–532.

la Cadena, de, Marisol., Indigenous Cosmopolitics in the Andes: Conceptual Reflections Beyond 'Politics', *Cultural Anthropology*, 25.2 (2010): 334–370.

Ezrahi, Y., *Imagined Democracies: Necessary Political Fictions* (Cambridge: Cambridge University Press, 2012).

Haraway, D., *Staying with the Trouble: Making Kin in the Chthulucene* (Durham, NC: Duke University Press, 2016).

Karvonen, A., and B. van Heur, 'Urban Laboratories: Experiments in Reworking Cities', *International Journal of Urban and Regional Research*, 38.2 (2014): 379–392.

Knox, H., 'Footprints in the City: Models, Materiality, and the Cultural Politics of Climate Change', *Anthropological Quarterly*, 87.2 (2014): 405–429.

Latour, B., 'A Well-Articulated Primatology. Reflexions of a Fellow-Traveller', in S. C. Strum and L. M. Fedigan, eds, *Primate Encounters: Models of Science, Gender, and Society* (Chicago: University of Chicago Press, 2000), pp. 358–381.

— *Facing Gaia: Eight Lectures on the New Climatic Regime* (London: Wiley, 2017).

Law, J., and E. Ruppert, *Modes of Knowing: Resources from the Baroque* (Manchester: Mattering Press, 2016).

Link, G., 'The Logical Analysis of Plurals and Mass Terms: A Lattice-Theoretical Approach', in P. Portner, and B. H. Partree, eds, *Formal Semantics: The Essential Readings* (Oxford: Blackwell, 2002), pp. 127–147.

Lovelock, J., *Gaia: A New Look at Life on Earth* (Oxford: Oxford University Press, 2000).

Marres, N., 'Issues Spark a Public into Being: A Key but Often Forgotten Point of the Lippmann-Dewey Debate', in B. Latour and P. Weibel, eds, *Making Things Public: Atmospheres of Democracy* (Cambridge, MA: MIT Press, 2005), pp. 208–217.

Morita, A., and C. B. Jensen, 'Delta Ontologies: Infrastructural Transformations in Southeast Asia', *Social Analysis*, 61.2 (2015): 118–133.

Povinelli, E. A., *Geontologies: A Requiem to Late Liberalism* (Durham, NC: Duke University Press, 2016).

Rancière, J., R. Bowlby, and D. Panagia, 'Ten Theses on Politics', *Theory & Event*, 5.3 (2001): n.p.

Rheinberger, H., 'Experimental Systems, Graphematic Spaces', in T. Lenoir, ed., *Inscribing Science: Scientific Texts and the Materiality of Communication* (Palo Alto: Stanford University Press, 1998), pp. 285–303.

Ruppert, E., 'The Governmental Topologies of Database Devices', *Theory, Culture & Society*, 29.4–5 (2012): 116–136.

Spencer, M., E. Dányi, and Y. Hayashi, 'Asymmetries and Climate Futures: Working with Waters in an Indigenous Australian Settlement', *Science, Technology & Human Values*, 44.5 (2019): 786–813.

Stengers, I., 'Comparison as a Matter of Concern', *Common Knowledge*, 17.1 (2011): 48–63.

— 'Gaia, the Urgency to Think (and Feel). The Thousand Names of Gaia' [from the Anthropocene to the Age of the Earth conference, Rio de Janeiro, 15–19 September 2014].

Thorpe, R. U., 'Democratic Politics in an Age of Mass Incarceration', in A. W. Dzur, I. Loader, and R. Sparks, eds, *Democratic Theory and Mass Incarceration* (New York: Oxford University Press, 2016), pp. 18–32.

Tsing, A. L., N. Bubandt, E. Gan, and H. A. Swanson, *Arts of Living on a Damaged Planet: Ghosts and Monsters of the Anthropocene* (Minneapolis: University of Minnesota Press, 2017).

Verran, H. 'Re-Imagining Land Ownership in Australia', *Postcolonial Studies: Culture, Politics, Economy*, 1.2 (1998): 237–254.

Verran, H., 'Number as an Inventive Frontier in Knowing and Working Australia's Water Resources', *Anthropological Theory*, 10.1–2 (2001): 171–178.

Wilkie, A., 'Prototyping as Event: Designing the Future of Obesity', in A. Corsín Jiménez, ed., *Prototyping Cultures, Art, Science, and Politics in Beta* (London: Routledge, 2017), pp. 95–111.

4

FIVE THESES ON ENERGY POLITIES

Brit Ross Winthereik, Stefan Helmreich, Damian O'Doherty, Mónica Amador-Jiménez, Noortje Marres

INTRODUCTION

Brit Ross Winthereik

IN HIS WORK ON CARBON DEMOCRACY, TIMOTHY MITCHELL (2011) RAISES A NUMBER of intriguing questions about the relation between fossil fuels and forms of governance. His main argument is that what he calls carbon democracies are geared towards maintaining and developing particular, fossil-based, carbon-intensive modes of energy production and distribution. The close interrelation between modes of governing and modes of fuelling societies in their current political forms suggests there is little chance that there will be any interruption or transformation of energy infrastructures in the near future. (We speak here mostly of the US, but Mitchell's point has been extended to other contexts.)

Andrew Barry (2013) has also analysed what he calls material politics as emerging in response to a particular concern with material infrastructure and natural resources – an oil pipeline, for instance. In his work we see how the process of forming public opinion about energy is varied and complicated. In comparison to Mitchell, who argues that infrastructures make public controversy around energy invisible, Barry's (2016) point is almost the opposite: infrastructures are drivers of political dispute, and even nature itself can play this 'unsettling' role.

What these authors show us is that modes of fuelling and modes of governing society can no longer be easily separated. What interests us here is whether and how renewable energy sources play a role in relation to governance and democracy. Still, today carbon is not the only source of energy, and political negotiation, contestation and struggle revolve around both fossil and renewable sources of energy such as wind, sun, biomass and water in the form of steam, rivers, tides or waves. As energy figures prominently in discussions around political power, governance, modes of knowing and organising society, the question remains how the processes through which energy becomes a matter of public concern actually happen. How do people engage with and through energy? What might the relations between energy infrastructures and democratic possibilities and impossibilities be?

The five contributing authors of this chapter attend to the political potential of particular energy sites. They show us how these respective energy sites afford very different forms of political potential, and have different effects on social and political gatherings. The question of whether an energy polity can be realised is the red thread that runs through the subsections of the chapter.

The chapter assembles stories of energy polities – emerging or fully formed social entities that draw political power directly from their engagement with environments where energy is at stake. The first section, by Stefan Helmreich addresses the question of sovereignty and the limits of political power at sites where nature and politics meet. Helmreich grapples with forces of arrogance and ignorance vis-à-vis natural forces and truth claims from democratic 'Others'. What results is a political battlefield as fraught with intensive force as the rising sea itself.

In the second section, Brit Ross Winthereik follows videos of wave energy technologies on social media. She asks whether the quantity of 'likes' and 'reshares' on these platforms signifies an emerging polity around wave energy, or inversely, whether this form of interest neutralises mobilisation around emerging energy technologies. Platform agency is conceptualised as anti-political due to its erasure of various forms of work and labour.

Section three, by Damian O'Doherty, takes the reader on a walk along a coastline estuary in England. The author and his fellow walkers contemplate if the clay from the landscape can be kindled into pottery, and reflect on the potential of this process as a storytelling mechanism that could provide an alternative, community-building map of this rich environment.

In contrast to the quiet contemplation during the estuary walking, Mónica Amador-Jiménez in her section invites us to consider the violent history of an oil-filled Colombian swamp (*ciénaga*). Here, she introduces us to different social groups living in and around the *ciénaga*. These groups are trying to non-antagonistically resist the lethality of the oil company and the paramilitary structures enveloping the *ciénaga*, as well as the state's abandonment of it. Amadór highlights the bio-sociality of the swamp as an entanglement of forces whose ontological interstices the inhabitants continue to negotiate in ways that seem to transgress 'normal' Colombian politics of left and right.

The last section, by Noortje Marres, takes us back to sites where technologies are being tested for their environmental and other capacities. Marres focuses on infrastructures as a topic of public interest, and specifically on the ways in which infrastructures draw in other times and places. For Marres, this has a democratic inflection. The various ways that infrastructures draw in the not-immediately-present makes us face problematic engagements that our understanding of democracy may not yet be equipped for.

In the context of this volume – where we consider energy as part of experimental socio-material worlds – the experiment performed in this chapter is to condense each analysis into a thesis (akin to the propositions in Chapter 3, but shorter). The five theses are presented at the end of the chapter on a 'scroll' that can be cut out and distributed, or framed to remind other publics or emerging polities of the democratic potential of energy. The theses are thus experimental devices, suggestive of possible democratic energy futures; they are instruments for speculation, but also themselves materials to work with. Each thesis revolves around the possibility that a polity may occur in relation to a particular site where energy features as part of an assemblage of nature, people, technologies and politics: a coastline, an online platform, a swamp, an estuary and a test site for driving.

The theses were developed through the authors' shared interest in the democratic potential of energy, and have been guided by questions such as: What if we imagined the relations between material environments and emerging political formations as laden with possibility? What insights emerge when we zoom in on particular socio-material sites where energy infrastructures are controversial and politically contested? What if we took the 'unfinished' nature of such sites as an analytical opening (Ballestero and Winthereik, 2021)?

The analytical site of intervention in our exploration is the family resemblance connecting a form of energy (carbon, for example) and a political form (democracy, for example). We want to defamiliarise this relation, as we see potential for change in reconsidering it. This potential is not to be considered some kind of promissory future, but rather a political form whose identity is not yet established. Described in our five theses are energy polities whose political action is under way, yet undecided. What follows is therefore a way of pondering multiple forms of political sociality in so-called post-carbon landscapes.

We hesitate to say post-carbon, because fossil fuels and renewable energy cannot be easily separated. Think about the oil and coal that goes into transporting giant wind turbines around the globe from production to installation. And consider political negotiations between actors that have come into being as an effect of, for example, installing wind turbines in a nature reserve. For us, the linkage of fossils and renewables speaks to an awareness of the many things that are configured together when we speak about the various forms politics might take at sites of energy.

In the tradition of philosophical thesis-writing (e.g. Marx's theses on Feuerbach, Benjamin's philosophical theses on history), the chapter strikes a critical-sceptical chord, a swan song for carbon-based societies. And the chapter shows that there is no carbon or post-carbon situation with 'natural', adjacent forms of political power. Instead, different situations embed different capacities for the formation of energy polities.

KING CNUT AND DONALD TRUMP AGAINST THE WAVES

Stefan Helmreich

In his *Historia Anglorum: The History of the English People*, penned in the twelfth century, Henry, Archdeacon of Huntingdon, included a homiletic narrative about the deeds of King Cnut the Great, a monarch (called 'Canute' in modern English) who in the early eleventh century had ruled over Denmark, Norway and England.[1] Henry's chronicle elaborated upon a legend in which King Cnut attempted to command the sea to cease its tides:

FIG. 4.1 Canute rebukes his courtiers, by Alphonse-Marie-Adolphe de Neuville

At the height of his ascendancy, he ordered his chair to be placed on the sea-shore as the tide was coming in. Then he said to the rising tide, 'You are subject to me, as the land on which I am sitting is mine, and no one has resisted my overlordship with impunity. I command you, therefore, not to rise on to my land, nor presume to wet the clothing or limbs of your master.' But the sea came up as usual, and disrespectfully drenched the king's feet and shins.

This story – sometimes known as 'Canute and the Waves' – has been employed by a range of commentators to describe the overreaching arrogance and ignorance of those in power, particularly when it comes to (under) estimating the forces of large-scale processes, including those considered 'nature' (Lord Raglan 1960). For example, in 2005 Louisiana lawyer Stacy Head slammed the New Orleans city council's feeble response to Hurricane Katrina by invoking Canute (Nolan 2009). Used in this way, the Cnut story is meant to point to the folly of seeking to control, in the realm of the 'political', energies that might rather belong to the domain of the natural (or, as we'll see in a moment, the supernatural). But according to University of Cambridge Professor of Anglo-Saxon, Norse and Celtic, Simon Keynes, the story is ultimately about Canute's wisdom, for Henry's tale concludes:

So, jumping back, the king cried, "Let all the world know that the power of kings is empty and worthless, and there is no king worthy of the name save Him by whose will heaven, earth and sea obey eternal laws." (Westcott 2011)

Fast forward now to the early twenty-first century – to a world in which it *is* possible, to some extent, to control and command ocean waves; to build infrastructures that protect shorelines, that 'harness' wave energy. As historians of surfing Peter Westwick and Peter Neushul (2013) have demonstrated in their book, *The World in a Curl*, waves have been created and destroyed around the world, sculpted in response to changing coastal infrastructures. And, as members of the 'Alien Energy' working group have shown, waves – in the form of 'wave energy' – have been eagerly enrolled by corporate and national initiatives into possible energy markets and polities. As waves become part of environmental infrastructures, and subject to new genres of political economic power, relations among the natural, the energetic and the political can now be imagined as synergetic (Helmreich 2016).

In October 2016, an international non-profit coalition called 'Save the Waves' called attention to a particularly potent contemporary political economic attempt to control waves:

> US President-Elect Donald Trump and his hotel company, Trump International Golf Links (TIGL), seek to build a massively controversial seawall on a public beach to protect his Trump Golf Resort in western Ireland.[2]

Here, Trump operates as the overreaching version of King Cnut, seeking to control the waves that might compromise an Irish golf course property he owns.

In the event, Save the Waves (which collected 100,000 signatures) successfully blocked the proposed wall, which was planned to 'run 2.8 kilometers, reach 15 feet tall, and consist of 200,000 tons of rock dumped in a sensitive coastal sand dune system'. But beneath this story is a weird wrinkle. Trump has famously dismissed the reality of climate change, but his organisation operates with climate change as part of its calculations and accounting. The environmental impact statement that Trump's people submitted in their original proposal for the Irish seawall read this way:

> *If the predictions of an increase in sea level rise as a result of global warming prove correct, however, it is likely that there will be a corresponding increase in coastal erosion rates not just in Doughmore Bay but around much of the coastline of Ireland. [...] The existing erosion rate will continue and worsen, due to sea level rise, in the next coming years, posing a real and immediate risk to most of the golf course frontage and assets.* (Sherlock 2016; my italics)

This is not a shift to the wise and humble version of King Cnut, recognising the limits of human sovereignty. The theory of sovereignty here is, rather, cynical – opportunistically using legal and scientific language without regard to the truth of the claims made, but prioritising instead the momentary rhetorical obfuscation such claims can enable.

ONLINE PLATFORMS AS ANTI-POLITY MACHINES

Brit Ross Winthereik

Above, Stefan Helmreich showed that an energy polity is made when NGOs form organised resistance to walls built on contradictory logics. But where else might energy polities take form? In cyberspace, perhaps?

Studies of the internet show that there was formerly much faith in cyberspace as a place for democratic deliberation and political mobilisation (Hague and Loader 1999; Jenkins and

Thorburn 2003). One could argue that what this faith in technology expressed was a hope for polity, and for establishing liberating organisational structures. The imagined power of the internet was precisely its capacity to facilitate the formation of political entities beyond any physical territory (Flichy 2007).

Today, twenty years on, the internet is something entirely different, in the sense that it has become a locus for surveillance capitalism (Zuboff 2019). Despite disappointments and scandals, however, there is still widespread belief that clicks and likes can work as a measure of actual public interest. The thesis on energy polity that I propose is based on an analysis of videos featuring wave energy technologies that I came across on YouTube and Facebook as part of fieldwork among wave energy inventors in Denmark.

Wavestar Energy is a Danish company that produces structures for the harnessing of ocean energy. Between 2009 and 2016 this company developed and tested several different technologies and, despite the many other Danish companies active in this area, it attracted significant financial investment. Having developed and tested many different prototypes, the company attracted enough money to be able to scale up one of these prototypes. For five years, a 1:2-scale wave energy device sat on the shoreline of north-west Denmark, while its technologies and capacities, including its capacity to withstand the power of the ocean, were being tested. Many visitors took this prototype to be the material proof of a growing marine renewable industry in Denmark, with sister industries around the Atlantic. Royalty, politicians, school children, businesspeople and tourists walked the steel bridge connecting the large steel construction to the shore. From there they would watch two enormous pontoons generating electricity from the movement of the waves.

Standing on the steel bridge, you would get a sense that what was given, in return for the journey to the north, was a glimpse into a green energy future. But standing there also allowed visitors to see something else: the continual rusting of the machine, and the human effort involved in maintenance, repair and data collection.

Although Wavestar demolished the prototype in 2017, this wasn't before it had been made the subject of a short video by an international visitor. He published it on Facebook, along with a text that claimed this to be the future of energy production. The video locates the Wavestar prototype, not only in Denmark, but also in a possible near future reality where energy production will come from waves. The video introduces the Wavestar prototype as an engineering wonder, as the camera zooms in on the hydraulic system and cuts to a panoramic scene and a scaled-up version of the device, centrally located among offshore wind turbines stretching as far as the eye can see. It gives the impression of peeping into a brave new world in which wave energy has become the norm. This impression is fortified by the final section of the video, which features a coastline (not Denmark, perhaps Southern Europe or Norway) where waves batter a rocky shore and the voice-over makes a statement: 'If we could capture just 0.1% of the ocean's kinetic energy, we could satisfy the global energy demand over 5 times'.

The video went viral. Today it has 54 million views, 900,000 shares and 12,000 comments.[3] It would not be that far-fetched to consider 12,000 comments as the expression of an energy polity – an entity without territorial claims, but with some sort of identity or voice. It is not an altogether outrageous thought to believe that a fraction of this vast number of views, shares and comments would be able to mobilise some kind of financial, public or political support for the company and the cause.

To find out whether this was indeed the case I talked to the inventor who, along with his brother, was also the founder of the company that later developed the Wavestar 1:2 scale prototype. Over lunch at the annual meeting for wave energy inventors in Denmark, he told me how surprised he had been by the popularity of the video. He also explained how he saw the video as having taken on a life of its own, a life separated from anything he knew about wave energy in his world of prototypes and engineering.

I have no real way of knowing whether clicks, likes and views on a social media platform can mobilise a polity around energy, but it seems there will be obstacles, which is why I offer the notion of the anti-polity machine. As an environment for the making of political subjects and subjectivities, social media renders crucial dimensions of wave energy invisible. In my view these dimensions are necessary to 'get' this energy form, its organisations and politics.

The video by Facebook user Hashem Al-Ghaili, described above, renders two things invisible. Firstly, it leaves out the human labour involved in the harnessing of wave energy. There is not a single person in sight in the footage, and energy conversion appears to be a smooth and clean process. Secondly, it presents oceans as undifferentiated masses of water; the same, no matter where one is on the planet. The video expresses a hope and an expectation, but, more than that, it works as an 'eraser'. Human labour is not present in the video, and, more than that, the forces of the ocean seem to have replaced human labour.

This is problematic for at least two reasons. Both waves and human labour must be considered as culturally, socially and technologically embedded for us to be able to understand the relation between humans and oceans (Helmreich 2014). The algorithmic production of wave energy on digital platforms codifies oceans while erasing any trace of the labour needed to realise this kind of renewable energy production. Not even millions and millions of views would be sufficient to summon up a polity around wave energy and a wave energy future. Social media algorithms are in place to ensure a seamless sharing of videos that are already popular. Analogous to James Ferguson's (1990) famous point that World Bank reports work as anti-politics machines in politically charged aid environments, I suggest we consider whether social media sites work as anti-polity machines – that is, as machinic environments that don't know how to differentiate between forms of labour, oceanic or otherwise, or describe their interrelation.

SCRYING HIDDEN ENERGIES IN THE DEE ESTUARY

Damian O'Doherty

How can we learn to hear the estuary speak? What does the estuarine landscape want to say? What is its business? These are the questions we are bouncing back and forth as we park up at Flint Castle car park on a bright and blustery spring morning in 2017. The castle was built in 1277 by Edward I, to facilitate the English military conquest of Wales. We read a sign from the Welsh tourist board that tells us that Flint castle was finally dismantled in 1652 and effectively 'buried in its own ruins' following the end of the English Civil War. These buried ruins loom large as we meet up with a group of local estuary artists we have arranged to follow for the day as they take a walk – or, as they call it a 'reccy' – around the tidal mudflats and the land which immediately borders the estuary. We have brought with us various recording and collection devices, including the ubiquitous ethnographer's bi-fold notebook and pencil, together with digital cameras and sound-recording equipment. Martin, Robin and Carl have brought a retinue of knapsacks and nets, waterproofs, waders and rubber boots, gloves, hats, binoculars and handheld GPS tracking devices. Against the backdrop of the ruins of Flint castle, we set off on our reccy, but only our equipment allows us to enter and navigate the treacherous and unstable territory where land meets sea in marshes and pools of brackish water left by the receding tide of the Irish Sea, pushed by the downstream egress of the River Dee.

We are immediately plunged into matter and movement, slipping, sliding, sinking, sucking, blowing, whistling air. We have to rely on our boots and the experience of the local artists as we pick our way across mudflats and sand, our hands and feet flailing for handholds and solid ground. As the tide recedes, wading birds are beginning to flock and gather on the exposed mudflats. Accompanied by an ambient soundtrack of shrill screeching, fluted whistles and liquid trills, we suddenly spot an oystercatcher flit across the skyline and land on a narrow spit of dark mud, the distinctive flash of its long bright orange beak deftly probing for food. Martin begins to sink his hands into what appears to be mud. But it is clay, we are told, as our hands explore its tacky wet-dry properties. It's a fine-grained mineral sponge made up in part of fibrous vegetable matter that squishes and stretches, sticks and rolls, holding its form while simultaneously shape-shifting with a plasticity that belies its appearance of slop.

A few moments later Carl comes to a halt and turns to approach the edge face of a low cliff wall. It is pock-marked and marbled with pebble-dashed ash-grey stone and coloured veins of burnished orange and cobalt blue, reminiscent of a Jackson Pollock 'action painting'. As Carl begins to explore the rock face, which in places reveals distinctive stratigraph-like layers of material residue, Martin begins to tell us that these are strange materials, probably all kinds of metals that have been compounded as aggregate and settled to form a hard, compressed 'rock' face. Around us on the floor are all manner of rock-like metals, blasted chinks of schist, and unidentified lumps of irregular and gnarled matter. We realise that this cliff edge fronting the estuary waters is the exposed face of the dismantled Courtaulds chemical works, which was located less than 100 yards from where we are standing; the rubble and spoil of the works was razed and levelled and then pushed to the fringes of the estuary and 'landscaped' to form a 'protective sea wall'. As we walk we become mindful that the ground beneath our feet is less solid than it might appear. We are walking on layers of time. Memories and stories begin to be shared amongst the walkers, of fathers who worked in the chemical works, of deep seam coal mining and subterranean gas pipelines still there below the surface. Martin begins to explore the clay again. The way in which the material resists is fascinating.

Looking up from the clay in my hands, I follow the contours of the land and wonder if the particulate matters in the clay carry residue of the materials and waste once used in the vast Courtaulds chemical works in the production of rayon, or 'fake silk' as Blanc (2016) has recently called it in his 'lethal history of viscose rayon'. Martin begins to show an interest in how this clay might fire in his kiln oven, a process which can apparently produce all manner of unanticipated blotching or streaks of colour that lend distinctiveness and idiosyncrasy to each piece of finished pottery. He explains that pots thrown from clay collected at different points in the walk might reveal subtle gradations in the accumulation of chemical and material composition: heavy metals, polychlorinated biphenyls (PCBs), hydrocarbons and other organic chemicals, all known to have been discharged into the Dee over the years. But what of the chemicals and toxicities not yet known, confessed to or recorded?

Martin picks up the discussion and wonders if a collection of pottery produced in this way could provide an alternative way of 'mapping' the estuary, which one might also consider a practice that *makes visible* or makes speak. Could the practices of storytelling involved in the intricate alchemy of pottery and kiln-firing tell intergenerational stories? For environmental author Barry Lopez (1998), 'kilns produce stories that emerge as part of a process', a process which later 'becomes visible in layers – layers of earth and fire, layers of emotion, ideas and change'. Most important for Lopez is what is found *between* these layers. It is this 'between' that demands most attention, but it is a between that lacks fixed 'bookends'. In something resembling alchemy, all that we might consider 'parts' are in flux (without an obvious 'whole'): the kiln and its materials, the potters and their 'skill', intentionality and design.[4]

For Lopez, the vibrations of these stories can be a resource for community building. Here, a collection of diverse potters (professional and amateur), children, backwoodsmen, artists, woodcutters, neighbours, the cussed and cockeyed hermits of the local wood, mix with the merely curious and passers-by. The kiln acts to stimulate the human community of potters into acts of generosity, sharing and humility. In another sense the kiln helps realise a recreational kiln-community, creating agency and proliferating relations in ways that give rise to new properties to which either the human or nonhuman might become attached.

But the shifting sands of the Dee estuary are a more 'dissonant' landscape, productive of 'danger, liminality and uncertainty' (Roberts 2016). In Martin's hands the Dee clay is disruptive, provoking memories and associations that lead from chemical toxicity to a history of coal mining and towards a possible future storytelling where the authorship becomes shared between human and nonhuman. We can think of these relations between physical/chemical matters and human matters as storytellers that help co-produce or animate what otherwise might lie dormant: matters of concern for the individuals and communities of the estuary that lie buried like the former coal mines, and physical or chemical matters that might otherwise be left unseen or unnoticed. Nonhuman partners help animate the estuary so that human-landscape entanglements are given the chance to realise their becoming-otherwise. In our hands, the clay draws attention to and animates the 'convulsive' landscape, the shifting sands and tidal energies, the grit that is 'between' the corrosion and the blending of elements, all placed within a vast temporal and spatial 'inhuman nature' (Clark 2011). Martin imagines a set of kiln-fired pots that will draw readers into the landscape and draw readings out of the landscape. 'How about we curate or put on display a set of Dee clay pottery?' Robin suggests. Robin considers the possibility of opening a 'pop-up' gallery, one that gives voice to the storytelling capacities of kiln-fired pottery and which might draw people in from the local communities. As we trawl the marshes and wetlands that border the estuary, we appear to be conspiring to imagine a new business start-up venture, but one that is also likely to make a political intervention. We stumble across a crop of wild samphire. Might we offer authentic estuary samphire tea, served in Dee clay pottery that also reclaims and renders inert old dangerous metals and chemicals? Stories shared over samphire tea are likely to be painful, however, especially those which recall the deaths and drownings inflicted by the dangerous tides of the estuary: lost fishermen and people last seen making their way across concealed fords guided by 'wayfinder' and 'ferrymen' who promise safe passage at low tide. In this way the community was also able to get rid of unwanted and troublesome elements, or so the myths tell us – debtors, adulterers or the unsuspecting Roman centurion or two coaxed away from Chester castle towards the Welsh border.

At some point we cross over the Welsh-English border – in fact we must have crossed back and forth several times. But a cartographic clue can be found in the observation made by Martin that he knows this land 'like the back of his hand'. At some point of obscure and contested origins the landscape becomes indiscernible from the human bodies which in part help bring forth the landscape through acts of shaping, ploughing, digging, building and so on. As it is worked, the landscape also works on its human workers. The furrows and lines on the hand of the potter are impressions formed by the furrows and lines of the clay taken from the landscape, but the landscape also takes back human bodies, which in turn help nourish and re-create landscape. Who is working who here? Who or what is the consumer/consumed?

Bodying forth in this space demands a form of narrative and description we are not used to in organisation studies. Ethnography offers perhaps our best chance; more specifically, it may be rewarding to develop the resources of this practice we have decided to call 'ethnogeomorphology'. With this concept we find ways of 'staying with the trouble' in Haraway's (2016) terms. Like Martin, tracking the land-water ways to find his way to the edge, we have to skirt carefully the language and concepts of the modern social sciences with which we have learned to make sense. Here, on the edge of the horizon, opposites meet and bleed into one another, but it is also where new divisions, distinctions and discriminations might be made. Things get put together differently here, opening up connections and relations that come prior to human, animal, sand, sea, samphire, zinc, polychlorinated biphenyls. This difference might demand a cultivation of what Derrida calls the 'chemical senses', which also seem to course through 'Bog', a remarkable 2018 essay by Mark Cocker. This essay draws on Seamus Heaney's understanding of a bog as a 'living, breathing organism'. Cocker asserts that Heaney's poetry 're-enacts the processes of transformation as body and bog coalesce'

(Cocker 2018: 242). In Cocker's own writing he not only *tells* us that the 'dark juices' of the bog '[work] upon the imagination in increments' but the writing itself enacts or becomes a living-breathing primordiality that takes on a life of its own for the reader (and, presumably, author): 'the breeze … the low-breathing bog, my own heartbeat … the songs of the birds swelled up to fill it and each of us was enlarged by the same creative process' (239). Land, bog, air, physico-chemical properties, sweat, breathing, writing – who is author-subject and what object? The narrative voice effects a separation of course, a distinction of sorts, but the sheer originality of the writing-bog has put the author together differently; the landscape allows 'new truths' to emerge for Cocker. Things are assembled here in ways that suggest the influence of an extended or liberated 'chemical sense' that is being turned to by a number of other writers as a way of harnessing existential transformation for the purposes of treating psychological ill-health such as anxiety and depression (Pollan 2019).

AMPHIBIAN POLITICS IN THE *CIÉNAGA*

Mónica Amador-Jiménez

An oil enclave is one of those places that we know exist, but are difficult to access for most people. Many of these enclaves are almost completely closed off to outsiders, while others are under a less restrictive access and security regime. Located far away from main cities, oil enclaves have become places of 'invisible' energopower that fuel societies (Boyer 2014). The oil enclave that is the focus of this paper, Velázquez oil field, is located in Colombia's Middle Magdalena Valley, and operated by the Sino-Indian oil company Mansarovar. This particular enclave is accessible, and is, moreover, inhabited by hundreds of people who have been living there for decades.

The oil enclave was built on top of a wetland system called a *ciénaga*. From the 1940s, the *ciénaga* has been made accessible by roads constructed to develop the oil enclave after the arrival of transnational oil corporation Texaco Petroleum Company, and it was soon populated by landless peasants fleeing political violence in in the north and south of Colombia.

In 1938, Texaco bought both the soil and subsoil of the territory where the enclave was established, based on a Royal Certification from the colonial Spanish crown. Therefore, this enclave formation must be understood not only as a residual effect of Spanish colonialism, but as the continuation of colonial power that persists in the forms of territorial organisation, land tenure and practices of social-racial organisation (Quijano 1992).

The *ciénaga* is located far away from Santa Fé de Bogotá, the capital of the Viceroyalty of Nueva Granada – as Colombia was called during colonialism – in a hot, flooded jungle. It was isolated until the eighteenth century, when the Spanish were able to access these territories. For two centuries, Maroons and Indigenous people encountered one other, and hiding together in the undergrowth they created a new world and a new ecology of practices that gave birth to a social group the Spaniards called *zambos*.[5]

A *zambo* was the offspring from the union of an indigenous woman and a black African slave. This new-born was, therefore, a non-slave, as the progeny of indigenous people were considered to be free. However, for the Spanish colonisers, a *zambo* was something grotesque, a rebel being who emerged from the interstices where the colony could not control the social and biological order of the new world.

Zambos born inside the watery territories of the middle Magdalena river had the strength and knowledge to navigate those turbulent, deceptive, dark waters plagued by crocodiles, snakes and mosquitoes. And paradoxically, as a free labour force they became the engine that mobilised the colonial and republican economy of Colombia until the twentieth century. The *ciénaga* is a place that has been shaped by those resistant to colonialism, those with the impetus to carry on despite wars and oil extraction – and yet these people were also in a paradoxical alliance with their exploiters. Thus, in the Middle Magdalena River, the first oil workers' union was formed in Colombia. It was followed by other trade unions that consisted of oil enclave workers, but also of fisherfolk, peasants, women, artisans and rowers – a socio-political organisation that emerged from the environmental and historical context of their conditions of production and reproduction.

In the 1970s, the *ciénaga*, which had been an ideal place to hide and resist colonialism, became a key area for drug traffickers. These traffickers operated with the tacit approval of Texaco, and formed paramilitary groups bent on minimising the effectiveness of the trade unions (Medina Gallego 1990; Molano 2009). These groups realised that the *ciénaga* was a strategically smart place to hide – fairly remote, yet at the same time a place where it was relatively easy to maintain logistical arrangements with sites where coca was cultivated and cocaine produced, with the Magdalena river providing a transport route connecting to the Caribbean Sea and onwards towards the United States.

The political ecology of drug trafficking has had severe effects on the surrounding area: the development of coca plantations, drug laboratories and money laundering schemes supported by illegal land purchases that transform forests into pastures for livestock. These combined elements create areas of parallel sovereignty. In this way, Texaco-extractivism and drug trafficking-paramilitarism have articulated in the low, hot lands of the *ciénaga* a particular type of moral topography. According to Taussig (2003), such places have been integral to Colombia from colonial times, since when the lowlands have always been considered brutal and savage, requiring illegality and violence to domesticate them, while the cold highlands are seen as civilised.

The period from the 1980s until the 2000s were particularly violent in the *ciénaga*. The paramilitary block of Puerto Boyacá turned the territory into a war laboratory that would become the 'Puerto Boyacá Model of Paramilitarism'.[6]

Without much intervention by the state, the *ciénaga* was gradually suffocated by the contamination of the oil company, the pollution from cocaine production and increased nutrient loadings from the expanding agriculture, the latter causing eutrophication, algae bloom and higher fish mortality. At the same time, those who resisted the paramilitary order also found death, and their mutilated bodies were often found floating in the murky, muddy waters of the *ciénaga*. A form of necropolitics governed this territory, catalysing the *ciénaga* into a miasmatic space of death, fear and horror (Mbembe 2003).

Over the centuries, inhabitants endured and formed a kin-community around and with this waterbody. As one of them asked me: 'where are we going to go when we have already been forcibly displaced'. The Colombian sociologist Orlando Fals Borda (1984) has stated that the inhabitants of the floodplains of Colombia are like amphibious beings who navigate the difficult climate, hiding during drought and re-emerging with the rains, moulding their lives according to the conditions. By their resilience in the face of environmental and socio-political hardship, these communities manifest an ontology of persistence.

With the inhabitants of the *ciénaga* I swam, sailed, washed my clothes, drank the water, ate the fish, immersed myself in the mud, felt the scorching sun on my skin, was bitten by the swarming mosquitoes. I was contaminated, but also free, which made me think of the *zambos*. The *ciénaga* impregnates the body of the inhabitants. It moulds them, inhabits them. The inhabitants make kin with the *ciénaga* in this watery land, which elicits an amphibious disposition (Haraway 2016).

Paramilitary demobilisation in Puerto Boyacá, the municipality where the *ciénaga* is located, began in 2003 and culminated in 2006. A total of 31,671 combatants were demobilised across the country, double the amount that the government and NGOs had expected. They had estimated that demobilisation would involve between 14,000 and 16,000 paramilitary personnel. Only 17,000 weapons were handed over to the government during demobilisation, the majority of them damaged or old. The former High Counsellor for Post-Conflict, Rafael Pardo, said, jokingly, that the paramilitary demobilisation was indeed a very peculiar demobilisation. Alvaro Villaraga has argued that it was not only full of contradictions and mistakes, but also that its lack of proper planning and follow-up made it inevitable that individuals and groups would rearm. Up to 20–30% of demobilised combatants had, after five years, re-enrolled in new paramilitary structures operating in the regions they had previously controlled. These emerging groups were under the command of middle-ranking paramilitaries who were not attracted by the benefits of demobilisation. These leaders established small and versatile military-criminal structures that incorporated demobilised paramilitaries who, for different reasons, had not participated in the demo-bilisation process, and gangs of young people from the slums, as temporarily contracted combatants.

But in spite of an incomplete demobilisation and recidivism, the dispersal of the para-militaries nevertheless generated a space, temporary, physical and ontological, in which the paramilitary forces, the bosses, did not govern the life and death of the people. This inter-stitial transition allowed for the re-emergence of the inhabitants and their socio-political organisations.

In December 2013, the inhabitants joined forces with the main trade union of the oil workers, the *Union Sindical Obrera*, to stage a well-organised protest against the company. According to rumours, ranchers and local Bacrims supported the protest by, amongst other things, donating meat (cows) to the protesters or by providing them with equipment and logistics to block the main road and access to the oil station (where oil is stored before being sent to refineries by truck).[7] As the protesters – the villagers, fisherfolk, oil work-ers and former paramilitaries – successfully blocked the compound offices, they literally paralysed the whole oilfield during the 15 days the protest lasted, generating economic losses for the company amounting to more than five million dollars. As Angela, the oil company's social worker recalled: 'They knew very well which were the weak points as they are all from here'.

The protesters had three main demands: 1) That the company respect the wage rates negotiated between the company and the trades union – an agreement that subcontrac-tors customarily refused to adhere to; 2) That the company intensify the clean-up of the *ciénaga* and expand the fish repopulation scheme; 3) That the internal dirt road of the enclave should be paved, as the constant traffic of heavy equipment on the road produces dust in the dry seasons causing respiratory diseases among the children living along the road.

It was a tragic event that triggered the protests. Just before Christmas, a female security guard, who lived in the *ciénaga*, died in a dramatic traffic accident on her way to work. The inhabitants directly attributed the fatal accident to the negligence of the oil company. This tragic and personalised backdrop seemed to have made the protests more radical, as there had been no killings or dramatic deaths in the *ciénaga* since the official paramilitary demobilisation in 2004. This episode revived the pain of the past in a different situation, one in which inhabitants felt free to protest and express their sense of injustice – something unthinkable before the paramilitary demobilisation.

The struggle was long, and it *was* a struggle. It was explicit: mobilisations, stoppages, road blockades, speeches, protests. Families even blocked the trucks that took the oil to the refineries. The unthinkable had happened in the land of origin of the paramilitary model; people protested as if they were living in the 1950s when unions had been strong. The heterogeneous inhabitants, with their paradoxical alliances in this abandoned land, succeeded in getting the road paved.

I am going to summarise some of the features of this amphibious politics to delineate my thesis of the elicitation of politics from the *ciénaga*: that it is a type of non-antagonistic resistance. It was a strategy of entangling and navigating other world-making practices[8], by circumnavigating the lethality of the oil company, the paramilitary structures and the active abandonment of the state. Nevertheless, there were also moments of antagonistic response, but at specific historical conjunctures. Such a strategy articulates contradictory poles, around which 'normal' politics in Colombia are organised, namely those between left and right. This amphibious politics entangles the forces of paramilitary groups, the oil workers, and the oil company's internal contradictions in its relations with social organisations. Together, these generate an ontological interstice through which the inhabitants can continue navigating the *ciénaga* in its difficulties and potentialities. It makes use of the knowledge practices that are entangled with more hegemonic knowledge infrastructures – such as the oil extraction industry – as a way of expanding, validating and protecting itself. It is deeply connected to the materiality of the place: a muddy, watery land; a cyclical environment of floods, rains, aquatic plants, sun and fish, from which a kin has emerged both physically and emotionally. And this material relationality elicits a social and political ontological practice, a material relationality that is generated by the sedimented coupling of substances like the mud, water fish, aquatic plants and even toxic discharges from oil production, as well as the rhythms of the seasonal variations and the efforts of the inhabitants to adjust to and intervene in these variations. These amphibian dispositions have been efficient in sustaining reality-making in the *ciénaga* until today, even in the most adverse conditions of contamination, war and states of exception[9]. The *ciénaga* and its inhabitants have persisted by expanding their dendritic networks, including through this text.

WHY WE MUST LOOK FOR THE DETERMINATIONS OF OUR ENERGY WORLDS BEYOND THEIR LIMITS

Noortje Marres

> *We may not experience our ignorance as such, but we are nonetheless ignorant.*
>
> DOROTHY E. SMITH (1987, 110)

To point to infrastructure as a site for the formation of publics is to challenge prevailing understandings of political democracy. There are several reasons for this. One is that infrastructures – from transport to utilities and communication – today present favoured objects and sites of *privatisation* across the world. It may seem that this removes infrastructure from the realm of public politics, even if the effect in practice may be the opposite: when infrastructures are taken out of state ownership and put in private hands, they are likely to become the subject of public disputes, about anything from land rights to lack of accountability (Barry 2013). However, if infrastructure complicates ideals of political democracy, it is also because, *from the perspective of lived experience*, infrastructure highlights our dependence on arrangements beyond our control. This becomes forcefully clear in the event of environmental disasters such as floods, or consumer scandals – the withdrawal of products that have been poisoned somewhere along a supply chain we barely knew existed (Guggenheim 2014). In such cases, in which infrastructure becomes a topic of public interest, it becomes apparent that we are more dependent on – more constrained by – others and elsewheres than we realised or wished, and indeed, *more dependent on others and elsewheres than many of our understandings of democracy allow for.*

For many, at least in the West, to talk about democracy is to invoke ideals of self-assertion. Here, to participate in publics is to express individual opinions and passions, to celebrate one's capacity for self-definition if not self-determination. Democracy, as feminists have long argued, has historically required the bracketing of material dependency; and the 'material production of everyday life' has been located in the private sphere, outside the domain of the public (Pateman 1989; Arendt 1958). However, those who have argued in favour of a more infrastructurally and materially aware understanding of the public have not necessarily managed to undo this habit of bracketing material dependencies and the material reproduction of everyday life in democratic thought, or even aimed to do so.

Historically, 'materialism' as a force in political thought and public mobilisation has favoured a select set of mostly industrial locations as sites where the politicisation of infrastructure is possible: for instance the factory, or the mine. In recent decades, a much broader range of sites of infrastructural politics has been studied by social researchers, including oil pipelines, urban transport, road building projects and recycling systems (Mitchell 2011; Tironi and Palacios 2016; Harvey and Knox 2015; Hawkins 2011.

These studies have highlighted how the building or transformation of a variety of infrastructures present an occasion for the political mobilisation of communities and the enactment of public controversy. Yet some of these studies uphold or reproduce the spatial biases of materialism, even if they have added environmental to industrial locations. They also tend to suggest that the requirements of the 'public-isation' of infrastructure can only be met in exceptional places, settings and environments – those equipped for scaling up the assembly of actors: the oil rig platform where unionised workers may attempt a strike, for instance, or the fracking location where a public encounter could be forced between critics and proponents of oil extraction operations. To be sure, this is in part an empirical matter. For instance, 'I' might think that a relevant site of infrastructural politics is the kitchen – a problematic site for the material reproduction of everyday life if ever there was one – but apparently this particular issue-complex is resistant to larger scale mobilisation under political banners. Still, shouldn't we ask what it is about our understanding of the public politics of infrastructure that leads us to re-produce gendered assumptions about the locations from which publics may be legitimately and effectively addressed?

This is partly why I have begun to study automotive technologies and their publics. Cars, and the car system, surely belong on the above list of prevailing societal infrastructures (such as transport, utilities). In recent decades they have also been subject to public mobilisation and placed at the centre of a whole range of political controversies – from air pollution to consumerism, and from labour rights to surveillance in the case of so-called 'smart' cars.

But cars equally instantiate the 'other' type of material entanglement: they present mundane devices of dependency in a most literal sense – in countries with crumbling transport infrastructures, the car is the device that materialises one's obligations to dependents visibly (and shamefully?) – ones relations of dependence with the young and the old who need transporting and cannot transport themselves. As such, cars present sites of *problematic entanglement*, in the technical sense of the term so lucidly offered by the feminist sociologist Dorothy Smith:

> The concept of problematic [...] directs attention to a possible set of questions that have yet to be posed or of puzzles that are not yet formulated as such but are latent in the actualities of our experienced worlds. [...] It responds to our actual ignorance of the

determinations of our local world as long as we look for them within their limits. [...] An inquiry defined by such a problematic addresses a problem of how we are related to the worlds we live in. We may not experience our ignorance as such, but we are nonetheless ignorant. (Smith 1987, 110)

Incidentally, it is this sense of problematic entanglement that I sorely miss when I hear public commentators and critics without any noticeable dependencies proclaim their critical judgements on 'the age of the car'. I am quite sure that, as long as environmentalism equates with the negation of the practical necessities by which most of us are bound, it will continue to be taken for the elitist concern that it today represents for too many. Could the formation of publics around cars be a way to explicate infrastructural dependency in its multiple senses; could these publics turn problematic entanglement into a public drama? As I have argued elsewhere, one of the features of problematic entanglements as a public and not a private form of association is that they jointly implicate relative strangers (Marres 2012). These material publics do not, as a matter of course, coincide with already existing social communities. Surely this places serious constraints on political mobilisation, although it also helps to clarify the problem with limiting material publics to a select set of industrial and environmental locations (the factory, the mine, or the transport network).

Could the idea that material publics only form in some exceptional locations – industrial, environmental – in part stem from a lack of interest in pursuing issue-associations that do not already map onto existing social and political relations (the community, the organisation)? Is it something to do with how we conceive of field studies of infrastructure, as requiring bounded situations? We may be beyond infrastructure denial, but many material publics remain bracketed all the same. Are we equipped to take seriously the possibility of infra-publics?

5 Theses on Energy Polities

Brit Ross Winthereik, Stefan Helmreich, Damian O'Doherty,
Mónica Amador-Jiménez, and Noortje Marres

Thesis 1

AN ENERGY POLITY *EMERGES* IN THE COMPLEX TRADING ZONE THAT ENROLS PHYSICAL FORCES, CONTESTED SOVEREIGNTIES OVER THOSE FORCES, AND COSMOPOLITICAL, ENVIRONMENTAL ACTIVISM.

Thesis 2

AN ENERGY POLITY *STRUGGLES* TO DEVELOP WHEN DIGITAL PLATFORMS DISPLACE THE LABOUR THAT IS NECESSARY TO GENERATE COMMITTED ENVIRONMENTAL RELATIONS.

Thesis 3

AN ENERGY POLITY *GESTATES* THROUGH CITIZEN-ARTIST COLLECTIVES THAT MAP WATER BODIES IN ARTISTIC FORM.

Thesis 4

AN ENERGY POLITY *CONGEALS* IN CONTESTED CARBON SATURATED BIO-SOCIAL SYSTEMS.

Thesis 5

AN ENERGY POLITY *NEEDS TO TRANSCEND* THE BOUNDARIES OF TRADITIONAL POLITICAL SPACES IF IT IS TO HARNESS THE POTENTIALS OF THE PUBLIC.

NOTES

1   See Henry, Archdeacon of Huntingdon, Historia Anglorum (D. Greenway ed., 1996).
2   See www.savethewaves.org/stoptrumpsirishwall/.
3   See www.facebook.com/ScienceNaturePage/videos/818130261652567.
4   The work of Strathern (1991) on fractals is suggestive here, as are recent references to the 'mereological' and mereology in and around the work of Latour, Stengers and Strathern. See for example Latour 2011.
5   Black African slaves who escaped slavery during Spanish colonialism in Colombia and hid in the jungles, where they formed communities called *Palenques*. For a discussion of 'ecology of practices' see Stengers 2005.
6   This paramilitary model was developed during the 1970s–1990s in the municipality of Puerto Boyacá where Ciénaga Palagua is located. From here it was promoted, exported and adapted to other parts of country, financed by drug trafficking money. Peasants were voluntarily or forcibly recruited and trained by Israeli and British mercenaries. This new type of paramilitarism is a central explanatory factor behind the extreme violence that has been observed in Colombia during the last thirty years or so, in which not only guerilla movements have been targeted, but also peasants, human rights defenders, social leaders, labour union activists and journalists.
7   Neo-paramilitary armed groups that are led by former paramilitary soldiers who were demobilised in 2004 or that never demobilised.
8   See, e.g., Tsing 2015; de la Cadena and Blaser 2018.
9   In Colombia, the notion of state of exception refers to a security regime that lasted from 1952 until 1994, originally introduced by the government of President Laureano Gómez to repress all political expressions other than those related to the Liberal and Conservative parties.

REFERENCES

Arendt, H., *The Human Condition* (Chicago: University of Chicago Press, 1958; repr. 1998).
Barry, A., *Material Politics: Disputes Along the Pipeline* (London: John Wiley & Sons, 2013).
— 'Infrastructures Made Public', *Limn* 7 (2016).
Blanc, P. D., *Fake Silk: The Lethal History of Viscose Rayon* (New Haven: Yale University Press, 2016).
Bowker, G. C., and S. L. Star, *Sorting Things Out: Classification and Its Consequences*, (Cambridge, MA: MIT Press, 1999).
Boyer, D., 'Energopower: An Introduction', *Anthropological Quarterly*, 87.2 (2014): 309–333.
Clark, N., *Inhuman Nature: Sociable Life on a Dynamic Planet* (London: Sage, 2011).
de la Cadena, M., and M. Blaser, eds, *A World of Many Worlds* (Durham, NC: Duke University Press, 2018).
Cocker, M., *Our Place: Can We Save Britain's Wildlife Before It Is Too Late* (London: Vintage, 2018).
Derrida, J., 'White Mythology: Metaphor in the Text of Philosophy', in J. Derrida, *Margins of Philosophy* (Chicago: University of Chicago Press, 1982), pp. 5–74.
Fals Borda, O., *Historia Doble de la Costa Tomo III Resistencia en el San Jorge* (Bogotá: El áncora, 1984).
Ferguson, J., *The Anti-Politics Machine: 'Development', Depoliticization, and Bureaucratic Power in Lesotho* (Cambridge: Cambridge University Press, 1990).

Flichy, P., *The Internet Imaginaire* (Cambridge, MA: MIT Press, 2007).

Guggenheim, M., 'Introduction: Disasters as Politics – Politics as Disasters', *The Sociological Review*, 62.1 (2014): 1–16.

Hague, B., and B. Loader, *Digital Democracy: Discourse and Decision Making in the Information Age* (London: Routledge, 1999).

Haraway, D., *Staying with the Trouble: Making Kin in the Chthulucene* (Durham, NC: Duke University Press, 2016).

Harvey, P., and H. Knox, *Roads: An Anthropology of Infrastructure and Expertise.* (Ithaca: Cornell University Press, 2015).

Hawkins, G., 'Packaging Water: Plastic Bottles as Market and Public Devices', *Economy and Society*, 40.4 (2011): 534–552.

Helmreich, S., 'Waves: An Anthropology of Scientific Things', *HAU: Journal of Ethnographic Theory*, 4.3 (2014): 265–284.

— 'How to Hide an Island', *New Geographies 08: Island* (2016): 82–87.

Henry, Archdeacon of Huntingdon, *Historia Anglorum: The History of the English People 1133–1155* (12th Century), ed. by D. Greenway (Oxford: Clarendon Press, 1996).

Jenkins, H., and D. Thorburn, *Democracy and New Media* (Cambridge, MA: MIT Press, 2003).

Latour, B., *What's the Story? Organizing as a Mode of Existence*, ed. by J. H. Passoth, B. Peuker and M. Schillmeier (New York and London: Routledge, 2011).

Lord Raglan, 'Canute and the Waves', *Man* 60.4 (1960): 7–8.

Lopez, B., *About This Life* (London: Harvill Press, 1998).

Marres, N., *Material Participation: Technology, the Environment and Everyday Publics* (Basingstoke: Palgrave, 2012).

Masco, J., 'Ubiquitous Elements' [presentation to 4S/EASST conference, Barcelona, 2016].

Mbembe, A., 'Necropolitics', *Public Culture*, 15.1 (2003): 11–40.

Medina Gallego, C., *Autodefensas, Paramilitares y Narcotraficantes en Colombia: origen, Desarrollo y Consolidacion, el caso de Puerto Boyaca* (Bogotá: Documentos Periodisticos, 1990).

Mitchell, T., *Carbon Democracy: Political Power in the Age of Oil* (London and New York: Verso, 2011).

Mol, A., *The Logic of Care: Health and The Problem of Patient Choice* (London: Routledge, 2008).

Molano, A., *En Medio Del Magdalena Medio* (Bogotá: Libro impreso, 2009).

Nolan, B., 'Stacy Head Has Rubbed Some People the Wrong Way, But Supporters Say Her Brash Style is Misunderstood', *The Times-Picayune*, 8 April 2009 <http://www.nola.com/news/index.ssf/2009/04/stacy_head_photo_for_terry.html> [accessed 28 August 2019].

Pollan, M., *How to Change your Mind* (London: Penguin, 2009).

Pateman, C., 'Feminist Critiques of the Public/Private Dichotomy', in *The Disorder of Women* (Stanford: Stanford University Press, 1989), pp. 118–140.

Quijano, A., 'Colonialidad Y Modernidad/Racionalidad', *Perú Indígena*, 13.29 (1992): 11–20.

Roberts, L., 'The Cestrian Book of the Dead', in *Literary Mapping in the Digital Age*, ed. by D. J. Cooper, C. Donaldson and P. Murrieta-Flores (London and New York: Routledge, 2016).

Save the Waves, 'Breaking News: Save the Waves and Irish Partners Defeat Trump's Irish Wall', *Save the Waves Coalition Blog*, 6 December 2016, <https://www.savethewaves.org/breaking-news-save-the-waves-and-irish-partners-defeat-trumps-irish-wall/> [accessed 28 August 2019].

Sherlock, R., 'Donald Trump Says Climate Change is a 'Hoax' But Tries to Protect His Irish Real-Estate

From Its Impact', *The Telegraph*, 25 May 2016, <http://www.telegraph.co.uk/news/2016/05/23/donald-trump-says-climate-change-is-a-hoax-but-tries-to-protect/> [accessed 28 August 2019].

Smith, D. E., *The Everyday World as Problematic: A Feminist Sociology* (Toronto: University of Toronto Press, 1987).

Stengers, I., 'Introductory Notes on an Ecology of Practices', *Cultural Studies Review*, 11.1 (2005): 183–196.

Strathern, M., *Partial Connections* (Savage, MD: Rowman & Littlefield, 1991).

Taussig, M., *My Cocaine Museum* (Chicago: The University of Chicago Press, 2003).

Timmermans, S., and M. Berg, *The Gold Standard: The Challenge of Evidence-Based Medicine* (Philadelphia: Temple University Press, 2003).

Tironi, M., and R. Palacios, 'Affects and Urban Infrastructures: Researching Users' Daily Experiences of Santiago de Chile's Transport System', *Emotion, Space and Society*, 21 (2016): 41–49.

Tsing, A. L., *The Mushroom at the End of the World: On the Possibility of Life in Capitalist Ruins* (Princeton, NJ: Princeton University Press, 2015).

Westwick, P., and P. Neushul, P., *The World in the Curl: An Unconventional History of Surfing* (New York: Crown Books, 2013).

Zuboff, S., *The Age of Surveillance Capitalism: The Fight for a Human Future at the New Frontier of Power* (New York: Public Affairs, 2019).

5

UNDA: A GRAPHIC NOVEL OF ENERGY ENCOUNTERS

Laura Watts, Cymene Howe, Geoffrey C. Bowker, with art by Neil Ford, lettering by Rob Jones

INTRODUCTION

Unda is inspired by the silence around words, and by the very different energy and meaning that images and text offer in graphic novel form. 'Silence around words' means both noticing the gap and whitespace around words and letters on the page, and the importance of what cannot be said, only alluded to. We are interested in imagining and knowing what cannot be spoken – and elusive, ethereal energy is our case study. We follow author Ursula Le Guin, who once wrote that she goes beyond explication through to a larger clarity by 'leaving around her words that area of silence, that empty space, in which other and further truths and perceptions can form' (Le Guin 2016: 50). How do we describe our experimental world with rigour when that world exceeds words, when experiences tend to be sensorially subtle, as with electricity? (Anusas and Ingold 2015)

Visual poems have long worked with their shape on the page, with the silence and revelatory meaning to be found in and around the text. The silence between words, where a different knowledge can be found and the academic argument transcended, is one aspect of our experiment. Our graphic novel uses words, but we invite readers to linger in the gaps between panels, between words and between lines. We know this requires a different kind of reading to make meaning. So we invite you to participate in our experiment to look and to feel this world, and become energised.

In this introduction we offer three guides. Our first guide is to how we wrote Unda, the infrastructure we engaged and the extensive invisible work we undertook to enact this graphic novel as a method. The second guide is to the format, to why we chose the graphic novel for our experiment, its role and histories. Here we offer some topical handholds for reading, should you want or need a 'way in' to our graphic novel as an academic experiment in energy research. Our final guide is to the themes in this graphic novel and the role of Unda herself as a figure of energy experimentation.

WRITING UNDA

During the workshop that led to this collected volume, the three of us were grouped together under the theme of 'interruption'. The short 'think pieces' we wrote in advance, and then discussed, interrupted standard modes of academic discourse and writing.

For Laura Watts (2019), this interruption registered through her juxtaposed poetic and narrative style – her figure of the Electric Nemesis can be seen swimming through a panel. For Cymene Howe (2019), her rich ethnographic observations in Mexico were constantly disturbed by the sound of [Red Tailed Hawk CAWS loudly] – a bird who is also an interrupting companion in our graphic novel. For Geoffrey Bowker, it was his attempt to look askance at the academic literature on energy infrastructure.

Tasked to trouble the waters of academic writing, we sought to counterpose insight and argument with experience and epiphany. Epiphanies – revelatory moments of knowing – interrupted in the way we sought.[1] As we developed our thoughts on publishing an interruption, it became clear that a mixture of text and image would be the revelatory format wanted. The graphic novel was a format that could provide a counterpoint to an argument, even as it attempted to draw a reader in. And so we began to consider how we would write such a novel in collaboration with a graphic artist.

It was a new form of engagement for us. We had to make a commitment to learning the graphic novelist's craft, which involved regular writers' meetings during a year-long collaboration, where we shared, explored and pushed our ideas. Let us just slow that down and point to the commitment, because successful experiments take time to develop and cannot be hurried. It would have been easier and quicker to turn the handle on a familiar writing practice, such as a standard academic paper, but that would have produced a familiar argument, a familiar energy world. So instead we held weekly video conference calls for a year. We talked from cars, from hotels, from flooded homes, from storms, in early mornings and late nights, across time zones that stretched from California to Europe. We established a collaborative workspace where we wrote and shared scanned images and inspirations.

We included our experienced graphic artist, Neil Ford, early in our development. Neil provided a well-established process for writing graphic novels that we could adapt to our own academic purpose. And, not to be overlooked, we found the necessary budget to support him as a freelance collaborator, and we honour him and our letterer (a separate craft in itself) as co-authors. We looked at examples of graphic novel scripts to learn the style and process. We had to move our thoughts on 'energy experiment' (the subject of our interruption) from concept to action, from theory to shapes on the page. Instead of an argument, we had to describe movement and motifs from one panel to the next. We had to think differently in order to write graphically – thus our arguments moved from being representation and metaphor to becoming embedded action, expressed in the description of a series of panels on a page. Often the most affective (and, we hope, revelatory) panels are those without words or letters.

The shift from metaphor to embedded action was not as straightforward and pithy as implied above, however. Since the graphic novel format will be new to many, we want to show the apparatus we used to construct our world – because apparatus and argument are inseparable in any experiment.

Here's an extract of an early script written for the scene, 'Encounter at Long Beach'. Perhaps four months into the process, we have already moved from our argument about the direct experience of energy and electromagnetic waves into motifs and emotional beats:

> It is evening. Unda has sailed down the coast toward Long Beach. She passes a Disney cruise ship with a large inflatable Mickey Mouse at the prow at the Port of San Pedro. She is blinded [by the lit-up cruise ship], horrified. There's something sinister about this cruise ship rearing over the stacks of containers in the harbor.

And, now, a later iteration of the script. You can see how we have had to go from general description to very specific actions so that the graphic artist can draw each distinct panel.

> PAGE 1 OF 3
>
> TITLE TEXT: Encounters at Long Beach…
>
> SETTING IS DAWN, SAN PEDRO BAY, OFF THE COAST OF PORT OF LONG BEACH, ONE OF BUSIEST CONTAINER PORTS IN THE WORLD. Sun has not yet risen. Unda has now sailed down the US West Coast to the Port of Long Beach. But she is just fleeing, not sure where she is going.

PANEL 1

Unda is headdown on the table, where we left her. We see a National Geographic magazine. And beside her head on the table is a smartphone that is bleeping. Through the portals is the looming front of a massive cruise ship – too close.

PHONE: Bleeeeep!

PANEL 2

View is over Unda's shoulder, as she looks down at her phone. It shows a nautical map with an outline of the San Pedro bay area, and a clear label at the top says 'Port of Long Beach CA'. There is a blob with a dotted line pointing at the ship symbol at the centre.

PHONE: Bleeeeep!

UNDA: Shit... collision alert. Where am I?

PANEL 3

Looking through two portals, through Unda's eyes: we see a vast Disney cruise ship that's blotting out the twilight sky. We can see the distinctive ears of a silhouetted Mickey Mouse at the fore, like a sinister monster looming. With the powerful lights of the cruise ship glowing like a halo around Mickey Mouse from behind.

When we wrote Unda, we wanted to have a lyric thread through our narrative, a mechanism to draw together our series of encounters. We began with music lyrics, but were thwarted by copyright issues. This led to a far deeper, oceanic connection with Samuel Taylor Coleridge's poem, *The Rime of the Ancient Mariner*. This classic poem contains an uncanny story of an energetic sea-borne being, which resonated for us and provided our lyric thread. We used the poem to weave our scenes and story together. It was only after completion that we discovered Nick Hayes' graphic novel, *The Rime of the Modern Mariner* (2012), which updates Coleridge to an ocean teeming with plastic and slick with pollution. Our work can be read in companionship with his.

Finally, one of our many important debates about apparatus concerned citation. How to maintain our commitment to academic relations? Techniques for remembering are well-honed in traditional academic texts: footnotes take us back and back through the echo chamber of more footnotes and references to the original source. But how do you cite and create footnotes within a graphic novel? How would that work visually on the page as integral to the experience we were crafting? Following academic custom, we could cite both graphics and texts, tracing the filiations back to an origin. For Foucault, the 'author' is not a person who originates a thought – this is an artefact of the copyright laws which followed hard on the heels of the invention of the printing press. Rather the author is a collective entity, drawing on all her senses simultaneously.[2] Scholars have invented a Procrustean bed for citation, but we were unsure if we should adopt it wholesale for our graphic novel, or whether emergent forms that are changing the very nature of philosophical argumentation should even engage with citation. Remembering has to be as multimodal as the object it endeavours to affix. So we offer this introduction as a way to perform some of our academic memories and filiations. And we invite readers to experiment with how to remember our graphical story.

READING UNDA

We chose the graphic novel as a format, not for its novelty but for its long literary heritage and recent resurgence, both within and without the academic realm. We were keen to learn how to write and be silent in this different literary mode, and to see what energetic worlds we could make that would have been impossible in academic prose.

Book chapters are a certain length because they often rest upon the unit of the journal article.[3] Lush pictures and graphic art have never been excluded from academic writing (witness the tradition in botany or zoology of finely drawn images), but in many fields the word is the key point of entry into the argument.

There are a number of existing graphical attempts to provide revelatory knowing beyond the limitations of the academic text. Our first inspiration was *Logicomix*, a graphic novel through which the reader meets the philosophers Whitehead, Russell and Wittgenstein (Papadatos et al. 2011). These three characters are typically troubled, passionate human beings – people caught in the swirling politics of their time; they are also depicted as talking heads who knew a logical conundrum when they saw one. There is no privileged point of entry into this graphic world; the protagonists are multi-layered and the graphic novel captures the instantaneity of their moments of understanding. Their assorted craziness is *part* of their logic, not a fleshy substance to wither away in the flame of their intelligence. Their story is told through words, scenery and graphical motifs, which weave together into meaning that is inherent rather than explanatory – meaning that is more akin to art than academic account.

Video ethnography and media studies are similarly fertile sites for visual rather than textual meaning-making, so we followed a further thread for our inspiration. Nick Sousanis' *Unflattening* (2015) is a graphic novel that engages with philosophical and social scientific discourse within science and technology studies. With a sparse and striking visual vocabulary (black-and-white line drawing) this graphic thesis weaves a rich narrative of souls unfurling as they come to appreciate the nested complexities of reality. A reader needs these visuals to appreciate the full body experience of the intellectual discovery of an Actor-Network Theory (ANT) informed vision of the world. The graphic novel panels are not illustrations of a theoretical point but are the unflat experiences that constitute the argument – the theoretical point is the experience.

We also read and studied other relevant works in the graphic novel genre: for example, Benoit Peeters' and Francois Schuiten's wonderful *Obscure Cities* series of graphic novels, which also occasions philosophical thought and critical thinking.[4] *My Favorite Thing is Monsters*, by Emil Ferris (2016), is the tale of a ten year-old in 1960s Chicago, with pulp horror movie styling. This marvellous creation speaks to the Holocaust and both echoes and transcends Art Spiegelman's better-known graphic novel, *Maus*. While *Maus* powerfully satirises the Holocaust in Poland – the Jews as mice, the Germans as cats – Ferris's work is a dark tale featuring monstrous synaesthesia: the monster learns to climb into paintings, to smell, taste and experience them with all her senses (Spiegelman 1986). Ferris's work evokes an orchestration of the senses, each thread of which must be savoured. To fully appreciate this work, the reader needs to spend as long on a graphic page as they would on a complex page of prose. Such graphic novels are not page-turners, easily digested, but the opposite.

Setting up the interplay between the visual and the literary has been a constant labour. Unda is not a linear story, even if it may appear as such in script or panels. Monsters splash through the pages in our examples above and the hallucinogenic images interrupt – as does Unda's own monstrous figure.

We see fiction as inherent to the work of writing an argument and to keeping alive the revelatory experience on the page. In this we resonate with Donna Haraway (2016), who has argued for the role of 'speculative fabulation' in scientific knowledge and world-making. Unda is an example of speculative fabulation – it is academic SF. Scholars have always used fiction in their critique and arguments: think of the interminable references to Funes the Memorious in works on memory; to the Garden of Forked Paths in the philosophy of quantum mechanics; and to citations of Borges' (2009) miscellaneous Chinese encyclopaedia in works on classification. Critical thinking and philosophical expression have never been limited to text.[5] Only when we exclude the fictional and visual do we fail to see the rich interplay between architecture, art, philosophy and science (Stafford 1994).

There is a conjunction between the form of representation and the form an argument takes. Academic writing, and the book chapter as a particular form of academic knowledge, is mutable (and *should* be mutable) – changing with the knowledge made.[6] Elizabeth Eisenstein (1979) has argued that absolute space and linear time were natural products of the medium of the book. However, the history of the book shows that non-linear and visual reasoning are as old as books themselves.

This can be seen in the transition from illuminated manuscripts to the textual book form.[7] Both intertwine text and graphic in order to tell the story or make the argument; illuminated manuscripts curl and weave their painted meaning around words and letters. So graphic reasoning came before our print age (and may outlive it). Graphic novels have retained techniques for circular and recursive reasoning. For example, Laurence Stern's classic *The Life and Opinions of Tristram Shandy*, an experimental novel published in 1759, is rich with graphics and has been characterised as a 'circular' novel (Douglas 2008). The importance of linear text may turn out to be a hiccup between times when text, image, and voice are threaded together into argument.[8]

Unda as a revelatory and reasoning graphic novel is only an interruption when contrasted with present academic forms of writing. Unda as an academic graphic novel is a continuation of visual experiments with a long pedigree. We are part of an old form of argument-making that we hope will be remembered this time.

WHO IS UNDA?

In this graphic novel we imagine a world that does critical work through its existence on the page. Perhaps the real difficulty with our graphic novel chapter is how to read it. We often see ourselves as *readers*: we read a map, a landscape, a text, a medical chart, a person's mind. Yet, *how to read* is the question, especially when the form is unfamiliar (Pound [1934] 1960). Social science and humanities scholars have been trained in the fine art of critical reading. We can take any academic paper, answer words with words, and weave our own story on the hypotext.[9] In the work that follows, we ask readers to learn to read the form of an academic argument as a graphic novel, even as we make our argument.

We can say things through the figurative representation of Unda and her relationship with her environment that we could not say otherwise. Some might feel we could have just said that we like renewable energy, and that we are not very happy with the continuing devastation of our planet through the stripping of energy stored over billennia (Mitchell 2013). However, we wanted to argue something different, with this emotive sense as backdrop. We wanted to see ourselves as energetic beings, transfixed in time, who resonate with, and draw from, the stored energy of the earth. We wanted to start to say that it's not just a binary between 'us' vs 'resources'; the hot vs the cold; honey vs ashes (Lévi-Strauss 1979). This work is an exploration of the deep continuity of our own being with the physical world. There is, we feel, a poetic meaning in the images – created from our script by Neil Ford, and lettered by Rob Jones – which complements the scarcity of words.

Unda, the protagonist of our story, is energy; she invokes energy and she encounters energy.[10] She is akin to Roy Scranton's (2015) *homo lux*, a being that embodies energy. We have tried to portray the deeply personal commitment the writers of this volume share towards energy experiments by celebrating the ways in which this works through our bodies and our actions, as well as through our words – on all levels at once. We remain committed to our empirical work and explicitly draw on our ethnographic research around energy infrastructures and their landscapes: Cymene's work in the Isthmus of Tehuantepec, Southern Mexico (Howe 2019; Howe and Boyer 2015); Laura Watts' work in Orkney, Scotland (Watts 2014 and 2019); the persistent threads of Geoffrey Bowker's (2008) commitment to infrastructure and its memory. These all provide a subtle fabric for the weaving of our graphic tapestry. We have learned that there is much that is invisible in energy, and that is where Unda lights our way. Invisible energy is at work in energy (not just electricity) infrastructures, the politics of power on our bodies, the unseen winds through which energy is made and transmitted. So Unda takes us from big, oily energy infrastructures, such as container ships, to the small particles that carry electromagnetic messages from satellites, to the energy that lights our cells and being within.

We imagine a subject (literally, a figure) who can sense energy, the full spectrum of electromagnetic radiation – light energy from gamma rays to radio waves – and she can feed off its power. Unda's journey takes the form of a *bildungsroman*, a novel about moral growth, that is mirrored in her body. She acquires new, untested abilities that seem overwhelming, and goes on a quest, as must we all, to sustain her vision in her darkest and lightest of times. We travel with her, hold on, as she sails through her electric tempest.

Unda invites you to join her energy experiment, to read between the words and the power lines, to see her energetic journey, and to find an altogether different energy world.

TIME TO GO...BUT WHERE? I NEED A PLAN. HAS TO BE SOMEWHERE DARK. AWAY FROM THIS SEARING ENERGY OVERLOAD...

MYSTERIOUS ORKNEY- WHERE'S THAT? AND IS IT DARK?

...60TH PARALLEL, NORTH OF SCOTLAND. SO, GOOD FOR HALF THE YEAR AND GOOD FOR NOW. I CAN WORK FROM THERE ...

*The Sun came up upon the left
Out of the sea came she!
Now wherefore stopp'st thou me?*

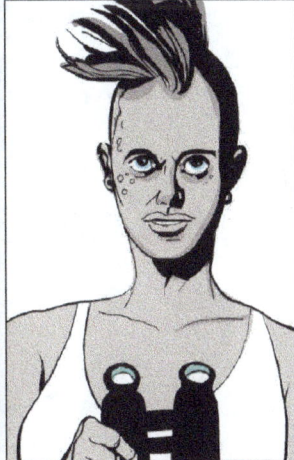

EVEN SATELLITES IN THE SKY HURT ME...THE PALEST ENERGY.

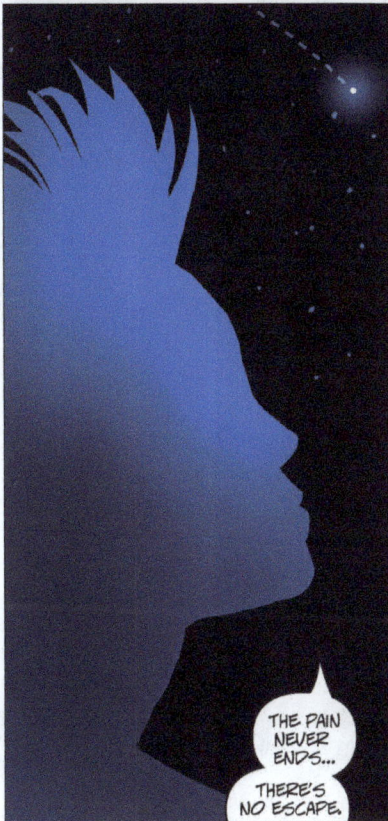

THE PAIN NEVER ENDS... THERE'S NO ESCAPE.

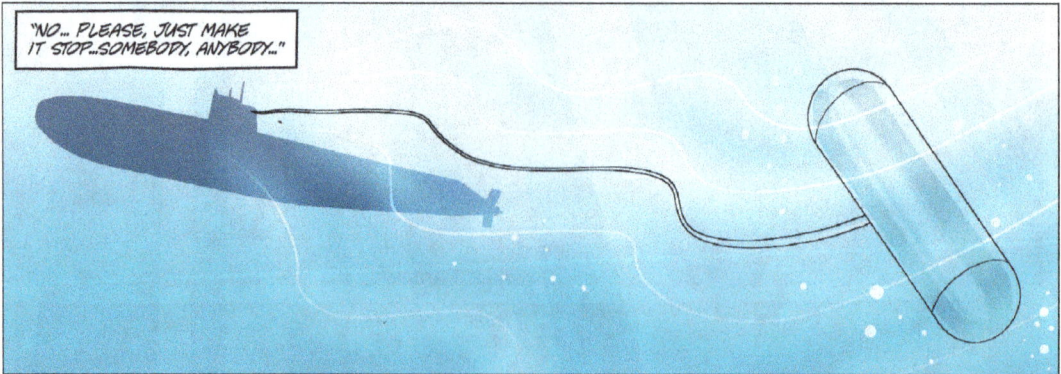

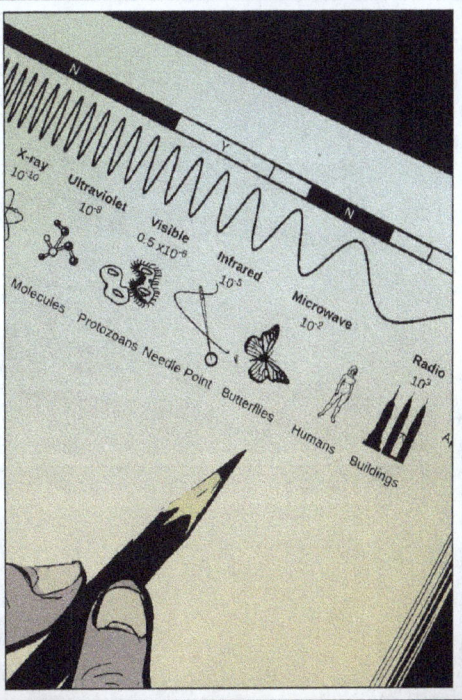

WAVEFRONT

Upon the whirl, where sank the ship
The boat spun round and round
And all was still, save that the hill
Was telling of the sound.

UNDA'S NEW LABORATORY...

"OTHERS MUST FEEL WHAT I CAN FEEL: THE INVISIBLE ELECTROMAGNETIC WORLD IN WHICH WE ALL WALK...LET US NOT BE OBLIVIOUS!

"I WANT OTHERS TO COME, TO HELP REBUILD MY EXPERIMENT, TO TRY IT THEMSELVES.

"I WANT PEOPLE TO KNOW WHAT HAS HAPPENED TO ME AND WHAT CAN HAPPEN THROUGH ME..."

THE ENERGY, IN THE AIR, IT BURNS ME...

...IT DRIVES ME TO SPEAK, TO BE HEARD, UNTIL OTHERS COME...

Since then, at an uncertain hour,
That agony returns:
And till my ghastly tale is told,
This heart within me burns.

NOTES

1   James Joyce writes beautifully on epiphany in *Ulysses,* as of course does Proust in *Remembrance of Things Past.*

2   See the lovely discussion in McGann 2014 *New Republic of Letters* of Poe as an archetypal collective author.

3   See Csiszar 2018 for a marvellous discussion of the development of the academic journal article.

4   See Peeters and Scuiten [1983] 2017. Other touchstones for us were the ancient, many tiered library in perpetual dissolution (L'Archiviste 2009) and the passion of the cartographer (La Frontière Invisible 2002). Each volume in the series invited us to think the world differently, askance from our normal vision.

5   With respect to art in critical argument, think of the rich analysis of Magritte's *Ceci N'est Pas Une Pipe* in Foucault 2010; or Foucault's marvellous reading of *Las Meninas* by Diego Velázquez (Foucault 1972); indeed, Louis Hjelmslev called Foucault 'a great audiovisual thinker' (Rajchman 1988).

6   For a discussion of the literary technology of academia see Shapin 1984.

7   Three examples of illuminated manuscripts are the notorious Codex Gigas (known as the Devil's Bible), which has a sensibility similar to Gothic comics today (Boldan, K., Kungliga Biblioteket & Národní knihovna České republiky. 2007; the eldritch Voynich manuscript, recently rendered in a beautiful facsimile edition (Clemens et al. 2016); and, as a later example, William Blake's *Songs of Innocence and Experience* (Blake [1794] 2004).

8   Kurke 2011 has a fascinating discussion of the context to the development of 'written' philosophy, and how it has become integral to current critical inquiry.

9   See Genette 1997. Hypotext is the subtending text, which is the foundation for an analysis, just as *The Odyssey* subtended *The Aeniad,* which both spoke to and echoed it.

10   Unda, as in undulation, suggesting waves. She is no relation to *Unda, The Bride of the Sea,* by H. P. Lovecraft.

REFERENCES

Anusas, M., and T. Ingold, 'The Charge against Electricity', *Cultural Anthropology,* 30.4 (2015): 540–54.

Blake, W., J. P. Witkin, J. Wood, and S. Albahari, *Songs of Innocence and Experience,* W. Slatkin (Ann Arbor: University of Michigan, [1794] 2004).

Boldan, K., *Codex Gigas – the Devil's Bible: The Secrets of the World's Largest Book* (Prague: National Library of the Czech Republic, 2007).

Borges, J. L., and A. Hurley, *Collected Fictions* (New York, NY: Penguin Books, 2009).

Bowker, G. C., *Memory Practices in the Sciences* (Cambridge, MA; London: MIT Press, 2008).

Clemens, R., D. E. Harkness, and Beinecke Rare Book and Manuscript Library, *The Voynich Manuscript* (New Haven: Yale University Press, [1401–1599?] 2016).

Csiszar, A., *The Scientific Journal: Authorship and the Politics of Knowledge in the Nineteenth Century* (Chicago: University of Chicago Press, 2018).

Douglas, M., *Thinking in Circles: An Essay on Ring Composition* (New Haven: Yale University Press, 2008).

Eisenstein, E. L., *The Printing Press as an Agent of Change: Communications and Cultural Transformations in Early Modern Europe* (Cambridge: Cambridge University Press, 1979).

Ferris, E., *My Favorite Thing is Monsters* (Seattle, WA: Fantagraphics Books, 2016).

Foucault, M., *The Archaeology of Knowledge; and, the Discourse on Language* (New York: Pantheon Books, 1972).

Foucault, M., and R. Magritte, *Ceci n'est pas une pipe* (Saint-Clément-de-Rivière, Hérault: Fata Morgana, 2010).

Genette, G., *Palimpsests: Literature in the Second Degree* (Lincoln: University of Nebraska Press, 1997).

Haraway, D., *Staying with the Trouble: Making Kin in the Chthulucene* (Durham, NC: Duke University Press Books, 2016).

Hayes, N., *The Rime of the Modern Mariner* (London: Viking, 2012).

Howe, C., *Ecologics: Wind and Power in the Anthropocene* (Durham, NC: Duke University Press, 2019).

Kurke, L., *Aesopic Conversations: Popular Tradition, Cultural Dialogue, and the Invention of Greek Prose* (Princeton: Princeton University Press, 2011).

Le Guin, U. K., *Words are my Matter: Writings about Life and Books, 2000–2016 with A Journal of a Writer's Week* (Easthampton, MA: Small Beer Press, 2016).

Lévi-Strauss, C., *From Honey to Ashes* (New York: Octagon Books, 1979).

McGann, J. J., *A New Republic of Letters: Memory and Scholarship in the Age of Digital Reproduction* (Cambridge, MA: Harvard University Press, 2014).

Mitchell, T., *Carbon Democracy: Political Power in the Age of Oil* (London: Vers, 2013).

Peeters, B., and F. Schuiten, *Samaris* (San Diego, CA: IDW Publishing, [1983] 2017).

Papadatos, A., A. Doxiadis, C. H. Papadimitriou, A. Di Donna, and M. Schifferstein, *Logicomix* (Amsterdam: Lebowski, 2011).

Pound, E., *ABC of Reading* (New York: New Directions, [1934] 1960).

Rajchman, J., 'Foucault's Art of Seeing', *October*, 44 (1988): 88–117.

Schuiten, F., and B. Peeters, *La frontiere invisible* (Bruxelles: Casterman, 2002).

— *L'archiviste* (Bruxelles: Casterman, 2009).

Scranton, R., *Learning to Die in the Anthropocene: Reflections on the End of a Civilization* (San Francisco, CA: City Lights Publishers, 2015).

Shapin, S., 'Pump and Circumstance: Robert Boyle's Literary Technology', *Social Studies of Science*, 14.4 (1984): 481–520.

Sousanis, N., *Unflattening* (Cambridge, MA: Harvard University Press, 2015)

Spiegelman, A., *Maus: A Survivor's Tale* (New York: Pantheon Books, 1986).

Stafford, B. M., *Artful Science: Enlightenment, Entertainment, and the Eclipse of Visual Education* (Cambridge, MA: MIT Press, 1994).

Watts, L., 'Liminal Futures: A Poem for Islands at the Edge', in J. Leach, and L. Wilson, eds, *Subversion, Conversion, Development: Cross-Cultural Knowledge Exchange and the Politics of Design* (Cambridge MA: MIT Press, 2014), pp. 19–38.

— *Energy at the End of the World: An Orkney Islands Saga* (Cambridge: MIT Press, 2019).

6

AN ENERGY EXPERIMENT: TESTS, TRIALS AND ELECTROTRUMPS

Jamie Cross, Simone Abram

POPULAR BIOGRAPHIES AND BIOPICS OF 'GREAT MEN' IN THE FIELD OF ELECTRICITY, such as Thomas Edison or Guglielmo Marconi, often suggest that the technologies we associate with them today were the inevitable outcomes of their endeavours. In some historical narratives it is as if a lightbulb or a radio were little more than an assemblage of conductive and insulating materials awaiting the ingenuity of a canny entrepreneur to realise their appropriate application. While anti-biographies, like David Nye's study of Edison, challenge these portraits of heroic individualism in electrical science and technology, there has been less attention to equally narrow depictions of the 'energy experiment'.[1]

Foundational scientific histories of electricity often hinge on dramatic experimental discoveries. Yet it is striking how little detail remains available of these events. Otherwise informative histories of early electrical experiments, such as that provided by André Koch Torres Assis (2010), include little technical detail on the scale or specificities of the experiments described. Instead, this book reduces such experiments to demonstrations of principle, perhaps highlighting the radical division between historical narratives of scientific development, and the more abstract or non-narrative academic communication of scientific findings found more commonly in journals. Similarly, studies of that challenging and quixotic subject, Nikola Tesla, put experimentation centre stage whilst embedding the experiment in a narrative of spectacular showmanship and even more spectacular failure (See Lomas 1999; Siefer 1996).

Like experiments in biology and biomedicine, however, experiments in the field of electricity invite questions about the production of scientific knowledge, as well as the changing relationship between science, technology and the public (Latour and Woolgar 1979; Latour 1993).

Our aim with this chapter is both to reclaim the specificities of electrical experiments and to develop a novel means of highlighting histories of experiment, introducing a simple game, 'ElectroTrumps', that sets up new opportunities for collectively engaging in juxtaposition, comparison and reflection.

WHEN IS AN ENERGY EXPERIMENT AN EXPERIMENT?

What is it that qualifies a practice as an 'experiment', as opposed to backyard tinkering? What kind of tests become widely recognised, and what kind of trials remain informal? 'Energy experiments' often fall along a *spectrum* of practices and methods for the production of scientific and technical knowledge.

At one end of the spectrum, we might say, is open experimentation that is designed to generate surprise, new knowledge and questions that could not have been asked before (Fischer 2007; Fortun 2014).

At the other end of the spectrum is the controlled test, designed to confirm what is already known. In between the two lies a glossary of terms (trials, demonstrations, probes, prototypes and pilots) that have distinct conceptual significance and ideological meaning in different contexts of use. For some, the opposite poles of this spectrum establish the distinction between two disciplinary traditions. As the pioneer of science and technology studies, Trevor Pinch (1993), once suggested, the test is to technologists what the experiment is to scientists.

Attempts to establish a precise typology – to pin down and anatomise the precise definitions of these words – would deflect attention from their fluid coexistence. In the contemporary world, from arenas of global finance to arenas of public policy, we find this spectrum of experimental terms deployed for particular ends – used strategically by people and institutions to claim particular qualities (novelty or originality, rigour and reliability, reach or generalisability) for their claims to truth.

EXPERIMENTS, ENERGY AND THE PUBLIC

What we refer to as energy 'experiments' might be said to represent a distinct genre of public engagement with the engineering and physical sciences. Over the past 250 years, experiments with the chemistry of materials and the organisation of energy systems have transformed public and private locations (from homes to museums, industrial laboratories, villages, towns, and cities) into truth spots for the production of scientific and technical knowledge. As the mode of energy experimentation has changed, so have the means by which materials, technologies and infrastructures are accepted or legitimised (Jasanoff 2003).

In nineteenth-century England, for example, energy experiments were performed in private, upper class stately homes. Such privileged venues have long given experiments authority and credibility – with vital implications for their reception by a wider, absent public (Shapin 1988). Cragside in Northumbria is one such instance: the home of Lord and Lady Armstrong was developed in the 1860s as a showpiece for hydroelectricity, with lighting, lifts and laundry services powered by the world's first hydro-electric power station.

Today, such high-profile energy experiments are just as likely to be held in public, or through public media, with scientists and engineers seeking to demonstrate the applicability of new knowledge by enlisting potential users and potential contexts of use into their experiments. Across Europe, for example, journalists now undertake 'green living experiments', setting themselves the task of living with low carbon or carbon neutral energy technologies for a fixed period of time and writing about their experiences in the print or online media (Marres 2008, 2009). Meanwhile television documentaries and current affairs programmes offer 'news' on new technologies, for example, as well as highlighting the promise and dangers of particular technologies, often with a tendency towards scandal or hyperbole, or a futuristic sci-fi slant to enhance perceived newsworthiness.

Just as biomedical technologies are granted credibility and authority through experiments that take place outside the scientific laboratory, experiments with energy materials technologies are made credible and authoritative by being conducted in these 'field labs'. In the UK, for example, publicly funded and visible trials of low carbon or energy efficient industrial processes – like the biodegradable waste treatment plants described by Joshua Reno – compel people to engage with energy science and technology in intimate and embodied ways (Kelly 2012; Reno 2010).

GLOBAL EXPERIMENTS

Rethinking moments of experimentation in the history of energy technologies also means re-distributing agency and authorship across global locations. The most important milestone in the twentieth century history of solar photovoltaics is usually recorded as the date in 1954 when Bell Laboratories publicly announced that three of its scientists – Daryl Chapin, Gerald Pearson and Calvin Fuller – had invented a silicon photovoltaic cell capable of converting enough of the sun's energy into power to run everyday electronic equipment. On 25 April 1954, the company held a press conference to announce the invention of the 'Bell Solar Battery', a panel of cells that could power a small toy, and the following day it presented to the battery to the National Academy of Sciences in Washington. *The New York Times* heralded the invention on its front page, writing that the solar cell

> […] may mark the beginning of a new era, leading eventually to the realisation of one of mankind's most cherished dreams – the harnessing of the almost limitless power of the sun for the uses of civilisation.

Yet this is a very particular account of the history of photovoltaic science and technology. It is a version of history in which agency is stabilised around three white men and one key material (silicon), rather than distributed across the complex network of humans and materials that were necessary for the solar cell to cohere as a successful technology. It is also a version of history in which agency is spatially located, with a North American scientific laboratory at the centre. Yet, as popular histories of solar energy reveal, non-western field sites and global locations have been critical sites of research, experimentation and testing in the development of the modern solar cell (Cross 2019).

By early 1954, Bell Laboratory's trials had led to the creation of a solar cell that crossed what was considered to be the minimum threshold for its viability, an efficiency of 6 %, producing 50 watts of electricity per square yard of photovoltaic material. Yet during the second half of the twentieth century, West Africa emerged as an important testing ground for photovoltaic applications in telecommunications; water pumping for drinking, livestock and irrigation; and lighting. One of the first systematically recorded 'experiments' to demonstrate a potential application for the modern, silicon solar photovoltaic cell in sub-Saharan Africa, for example, took place in Mali in the 1970s.

As John Perlin (1999) has documented, one French missionary, Bernard Verspieren, responded to the effects of a devastating drought by installing a solar powered water pump in a Malian village on the edge of the Sahel. Launching the solar powered pump, Verspieren told assembled villages: 'Solar power is the answer, it will be your salvation. You've seen it, touched it, listened to it. Not in a laboratory but in your own backyard.'

Over the next few years, Verspieren's detailed technical reports provided a stimulus to discussions about the coating of solar modules, and engineers responded by developing a more rugged design and more durable moulded glass panel, which more completely sealed the cells and their connections from contaminants such as dust and sand. This is but one example of the ways in which scientific and technical knowledge critical to the development of the silicon based solar photovoltaic module was produced, not in the laboratory spaces of Europe and North America, but in field laboratories across the non-Western world.

THE ELECTROTRUMPS GAME

Our reflections on energy experiments and public engagement led us to develop our own experiment in research, writing and play. Inspired by Joseph Dumit's (2017) innovation with learning-games, as well as Anna Tsing and Elizabeth Pollman's (2005) multi-round, 'Global Futures' game of imagination and speculation, we developed a card game about energy experiments. Our aim was to illustrate the wide range of experimental and test-like activities that have paved the development of electrical sciences and technologies. How would it be possible to classify these experiment-tests? What would be the key features to compare, and how would one decide which to include in a history of electrical experimentation?

Our game takes its inspiration from a popular cultural gaming phenomenon called Top Trumps. Top Trumps was first launched in the UK in 1968, and over the past fifty years hundreds of versions of the game have been produced.[2] By the 1980s the game was ubiquitous in primary schools, and the cards continue to be a popular staple of school playgrounds in the UK. The original cards were designed to promote basic literacy, numeracy and mnemonic skills as well as general knowledge – expressing a commitment to learning through play.

FIG. 6.1 Top Trumps Series 2, 'Prototypes'

Each pack of 52 Top Trumps cards is based on a theme or topic, from vehicles to space phenomena, from dinosaurs to wildlife, from movie stars to athletes. Each card presents a person, place or object linked to the theme, and a list of between four and eight vital attributes, which are each given a numerical value. The aim of the game is to win all the cards, and each round is played by a player choosing one of the categories as its 'trump'. The card with the highest value in this trump category beats all others, and the player with the winning card gets all the other cards played in this round, and goes on to choose the next trump from the next card in their pile. The classic game is more complicated than 'Snap' but far simpler than the multi-round imaginative storytelling card game designed by Tsing and Pollman.

Our modified version of the game, which we've called ElectroTrumps, has an energy experiments theme. The cards are intended to highlight the range and variety of energy experiments, tests and trials. In developing a template for the ElectroTrumps cards we settled on five categories: 1) Name (or author) of experiment; 2) Date of trial; 3) Voltage achieved in test; 4) Current used in test; 5) Significance of discovery from 1–10.

ElectroTrumps is based around the notion of categorisation, one that is central to anthropological theory. Social categorisation is the basis of most kinds of thought and action that anthropologists study, and the question of how to categorise is a basic human puzzle. Top Trumps can be understood as a way to learn about categorisation through play. The very definite and hierarchical form of categorisation used here, and its deployment in a competition (the player with most cards has won the game), reflects a distinctive socio-historical moment that chimes well with the history of electrical developments.

Our choice of categories is not entirely arbitrary, but neither is it definitive. Factors such as the date of the experiment and its power may or may not be established. The 'significance' of an experiment is largely debatable – subject to claims and context. This is partly the point of the game – to discuss the relative significance of different experiments, potentially prompting players to seek further information or debate the relative merits of different technologies.

Rather than compile a complete set of cards for distribution, we developed a small set of sample cards. Part of the play in this game is that players are invited to invent their own cards. Our sample cards are intended to inspire contributions from others. We propose that ElectroTrumps be played as a kind of wiki-game, where the fundamentals of the game might also find new forms or inspire new sets. Our aim is to propose a sample of cards that might inspire others to seek out further examples, explore the history of electrical play and reflect on choices about categorisation. Part of the game is the completion of a pack.

As currently envisaged, the game can encompass foundational experiments in the history of electricity – from well-known experiments like those carried out by Faraday, Edison, Swan and Tesla – to contemporary experiments with smart, digital systems, or low-tech renewable mini-grids. Our initial set of examples includes both formal, historically evidenced, laboratory experiments, and public-science tests. A set of definitions and guidelines related to the categories is provided, but is open to challenge and could be redefined in play.

The game has been designed for use as part of a classroom teaching activity, and aspects of the game were developed during seminars with undergraduate and graduate students at the Universities of Edinburgh and Durham. In making these suggestions, and by opening the game to direct input, we also propose the game as an alternative, experimental mode of teaching and public engagement with the humanities and social sciences – one that holds out the possibility of generating new and surprising questions about the social and material politics of energy within and beyond the academy.

The game is hosted on a dedicated website that contains a downloadable selection of cards as well as resources and a template for inventing cards (www.electrotrumps.xyz). Our guidelines for the game are not identical with the rules provided by the official, branded Top Trumps games. The materials are open to alternative uses, indicating the generative potential in play as a way of bringing forward alternate histories of electrical technologies.

In designing the cards, we adopted the principles of Top Trump card design, but then adapted these principles to our needs. We included a space for an image of the chosen experiment, which is named and located geographically, and five defining categories: the date of the experiment or test, whether the product used renewable energy, how dangerous it was, whether it could be considered successful, and how significant it was. The latter three categories are evaluated on a scale of 1–10, but the scores given are relatively arbitrary; it remains a matter of opinion how dangerous, successful or significant they may be. On the back of each card is a short summary of the experiment in question – a brief narrative account that helps to contextualise and explain the principles being tested. Writing these narratives is a challenging task, requiring the author to summarise a complex event or process in very few words. This, in turn, requires considerable reflection around the significance of the experiment, which in turn can be helpful for attributing scores in each category. What becomes clear from the original set of cards is that the format is surprisingly flexible, accommodating events as diverse as the 1794 Alessandro Volta battery and the 2008 Solar Suitcase in Abuja. This very diversity should help to prompt questions in the players' minds about what constitutes an electrical experiment and what the diverse examples in the pack have in common.

The following rules and questions demonstrate these aims, gently opening up the possibility not only to play the game as given, but to play with the game itself. Shall we play with these cards, or make our own? Do we accept the categories proposed by the inventors, or do we have better ones? Should there be more categories or different ones? What makes one experiment better than another? Which category is more important in which context? We hope that by presenting this starter-pack, we might ignite interest in electrical experimentation, and attract attention to the peculiarities of electrical technologies. By including specifically social categories (significance, success, danger), we hope to raise debate about what electrical technologies are, and their role as socio-material objects and practices in the world. We hope, also, to offer a bridge between engineering and social sciences, where playing together can open up debates and jokes for exploring ideas about the social and the technical.

RULES FOR PLAYING ELECTROTRUMPS

ElectroTrumps is a game for two or more players. The object of ElectroTrumps is to complete a full pack of playing cards and then to win all the cards. ElectroTrumps can be played with a deck of 30, 42, 48 or 52 cards. A selection of 15 ElectroTrump cards can be cut out from this book or downloaded from www.electrotrumps.xyz. Additional cards must be designed by the players. Players can use or modify an ElectroTrumps playing card template – which can be reproduced from this book or downloaded – to invent new cards. In the process of inventing new cards, players are invited to discuss or modify the theme, develop sub-sets, and review the categories.

To play, begin by shuffling a full deck of playing cards. Deal the cards until each player has an equal number, discarding any remaining cards. Agree who will begin, and decide the direction of play. The starting player chooses one of the categories or attributes on the cards as a trump. One at a time, each player selects one card from their hand and reads out the value of that category or attribute. Whoever has the highest value in the trump category wins, and collects all the other players' cards in this round. The winning player chooses the next trump category from another card in their pile. If a player loses all their cards, they are out of the game. The rounds continue until one player has collected all the cards.

There are two alternative modes of play. In the first, 'The Electron Variation', players can view all the cards in their hand and choose between their cards in each round. The winner can view their new cards. In the second, 'Insulator', players hold their cards face down so they cannot be seen; in each round they must choose the first card in the stack. The winner must place any new cards at the bottom of the stack, face down.

QUESTIONS TO CONSIDER DURING PLAY

- How important is it that the power-ratings are included in the experimental details?
- Is it fair to prioritise power-rating over inventiveness?
- Would it make more sense to choose categories that reflect the cost of experiment or the bravery of the authors of the experiment?
- Should there be a danger rating or a social-usefulness rating?
- Should there be a category for pollution or waste resulting from the test? Or a time-range for how long the test was relevant?
- Should we include a category of test/trial/experiment, and if so, what should be their hierarchy?
- Should the 'author' of the experiment always be included, or any form of identification be allowable?
- Does an older date trump a newer one, or vice versa? Or should this be negotiated at the start of each play?
- How should we rate the significance of the examples, or should we allow for a completely arbitrary scale based on our own judgements?
- How do we define the electric-ness of the experiment, and should it include any experiment that involves electric current, static or voltage?
- Will the cards highlight gender disparities in the power industries and in engineering research, or will they uncover hitherto unrecognised figures?
- Will they demonstrate an emphasis on particular fields, or reveal a funding-driven focus on particular technologies?
- Could they be brought together to design new experiments, invent new technologies or bring creativity to bear on under-emphasised challenges?

Download the full set of cards at matteringpress.org or electrotrumps.xyz to make your own Electro Trump Playing Cards.

ZETA FUSION REACTOR
OXFORDSHIRE, ENGLAND

DATE	1958
RENEWABLE	NO
DANGER	7
SUCCESS	3
SIGNIFICANCE	7

Zeta Fusion Reactor (Front)

Known at the time as "Britain's Sputnik" this experiment sought to develop Nuclear Fusion through the Z-Pinch technique. Carried out against the backdrop of Britain's decline as a world power and the Suez Crisis, this experiment signaled a shift in the way science and technology were deployed to exert global influence. The press hype did not match the end result, and was seen as a PR disaster – although the experiment laid foundations for important developments in nuclear energy to come.

Zeta Fusion Reactor (Back)

FARADAY'S INDUCTION EXPERIMENT
ENGLAND, HERTFORDSHIRE

DATE	1831
RENEWABLE	NO
DANGER	1
SUCCESS	9
SIGNIFICANCE	1

Faraday's Induction Experiment (Front)

ELECTRO TRUMPS

Using two wires wrapped
around an iron ring, Faraday noticed
that when he repeatedly plugged
and unplugged one wire into a
battery, attaching a galvanometer to
the other wire, it produced a "wave
of electricity." This was caused
by a change in magnetic flux.
It is this fundamental operating
principle that many generators,
engines and inductors rely on.

Faraday's Induction Experiment (Back)

LUIGI GALVANI'S TWITCHING FROG'S LEG EXPERIMENT
BOLOGNA, ITALY

DATE	1794
RENEWABLE	NO
DANGER	1
SUCCESS	3
SIGNIFICANCE	5

Galvani's Twitching Frog's Leg Experiment (Front)

ELECTRO ⚡ TRUMPS

In 1780, the physician, physicist, biologist and philosopher Luigi Galvani was dissecting a frog's leg when his steel scalpel accidentally brushed the brass hook holding the amphibian's leg in place. The leg twitched. Galvani believed that he had discovered what he coined "animal electricity" – the life force that resides in the muscles of animals. This theory would later be discarded by Galvani's friend, and rival, Alessandro Volta.

Galvani's Twitching Frog's Leg Experiment (Back)

ALESSANDRO VOLTA AND
THE FIRST BATTERY
PAVIA, ITALY

DATE	1794
RENEWABLE	NO
DANGER	1
SUCCESS	10
SIGNIFICANCE	9

Volta and the First Battery (Front)

Although the term was coined by Benjamin Franklin, it is Volta who is credited with the invention of the first battery. Volta's disputed Galvani's animal electricity theory, theorising that the twitching frog's leg was merely responding to the electricity - that a current was caused by the contact of two dissimilar metals in moisture Although the two were friends, their supporters clashed in the street over the two different theories. The "volt" was officially established in 1881 as a unit of electrical measurement in honour of Volta.

Volta and the First Battery (Back)

generalfusion®

GENERAL FUSION
BRITISH COLUMBIA, CANADA

DATE	2002 - PRESENT
RENEWABLE	YES
DANGER	10
SUCCESS	5
SIGNIFICANCE	9

General Fusion (Front)

Established in 2002, this crowdfunded project sought to use magnetized target fusion to create the most practical and quickest path to commercial nuclear fusion. It is an example of how it is no longer just national governments that fund large-scale scientific experiments.

General Fusion (Back)

**BENJAMIN FRANKLIN AND
THE KITE EXPERIMENT**
PHILADELPHIA, USA

DATE	1752
RENEWABLE	YES
DANGER	10
SUCCESS	5
SIGNIFICANCE	9

Franklin and the Kite Experiment (Front)

Proposed, and possibly carried out, by Benjamin Franklin the "kite experiment" has become one of the most widely known experiments in the world. The experiment itself, however, did not actually advance scientific knowledge. Yet, due to its mythical nature (the inherent danger of the experiment, the elemental force of lighting and the fact that Franklin was a Founding Father), it constructed a view of science in which progress is made through moments of individual brilliance.

Franklin and the Kite Experiment (Back)

COLVILLE LAKE SOLAR PROJECT NORTH WEST TERRITORIES, CANADA	
DATE	2013 - PRESENT
RENEWABLE	YES
DANGER	3
SUCCESS	7
SIGNIFICANCE	8

Colville Lake Solar Project (Front)

The people of the Arctic face the challenge of diversifying their energy sources. Due to poor infrastructure and harsh climate, communities in the Canadian Arctic use twice as much electricity as the national average. The Colville Lake project is the first renewable system in the North West Territories benefiting the First Nation communities that live in the vicinity, especially in the summer months. However, solar currently still costs more than diesel due to the cost of the batteries needed to store excess power.

Colville Lake Solar Project (Back)

RENEWABLE ENERGY AND GENDER
IN RURAL COMMUNITIES OF
NORTH-WEST CHINA

DATE	2014
RENEWABLE	YES
DANGER	2
SUCCESS	8
SIGNIFICANCE	6

Renewable Energy and Gender (Front)

ELECTRO TRUMPS

Many households in rural China rely on biomass-based energy. Collecting firewood is labour intensive - a task normally carried out by the women of the village. This study showed that by switching to renewable energy sources, women could save time and money giving them more freedom to pursue other activities. The switch to a cleaner fuel also has positive health effects as indoor smoke pollution can cause serious respiratory issues. The study shows how gender is an important, but often under-researched, dynamic of energy consumption.

Renewable Energy and Gender (Back)

**PERTH WAVE
ENERGY PROJECT**
PERTH, AUSTRALIA

DATE	2014
RENEWABLE	YES
DANGER	3
SUCCESS	7
SIGNIFICANCE	5

Perth Wave Energy Project (Front)

This experiment uses submerged buoys to harness the tidal forces readily available in Australia. The experiment is the first largescale commercial use of buoys as a source of energy (rather than panels), theoretically, requiring less upkeep. It could present a viable energy alternative for island nations who's future energy security is increasingly uncertain.

Perth Wave Energy Project (Back)

MANHATTAN PROJECT/TRINITY TEST	
NEW MEXICO, USA	
DATE	1942-1945
RENEWABLE	NO
DANGER	10
SUCCESS	9
SIGNIFICANCE	10

Manhattan Project/Trinity Test (Front)

Arguably one of the most significant and far-reaching scientific experiments ever, the Manhattan project saw the development of the first nuclear bomb, which was tested in an isolated corner of the New Mexican desert on July 16, 1945. Its development paved the way for nuclear energy as well as fundamentally changing the nature of warfare, setting the geopolitical stage for the century to come.

Manhattan Project/Trinity Test (Back)

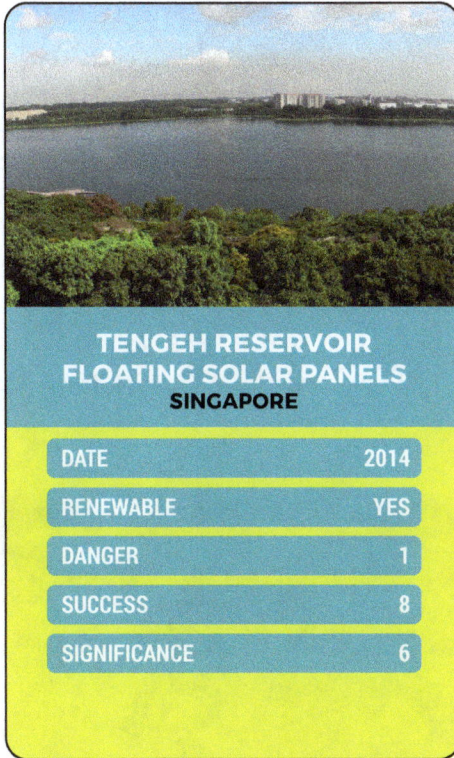

**TENGEH RESERVOIR
FLOATING SOLAR PANELS**
SINGAPORE

DATE	2014
RENEWABLE	YES
DANGER	1
SUCCESS	8
SIGNIFICANCE	6

Tengeh Reservoir Floating Solar Panels (Front)

ELECTRO ⚡ TRUMPS

Land constraints can be a significant obstacle in transitioning to solar power. The Tengeh Reservoir Solar Test Site is the largest floating solar photovoltaic cell test-bed in the world. The reservoir allows for easier cooling as well as allowing Singapore to build its solar capacity despite its dense urban environment.

Tengeh Reservoir Floating Solar Panels (Back)

THE TRANSACTIVE GRID
PRESIDENT STREET,
BROOKLYN, NEW YORK

DATE	2016
RENEWABLE	YES
DANGER	0
SUCCESS	8
SIGNIFICANCE	10

The Transactive Grid (Front)

This is widely cited as the world's first blockchain based, peer to peer energy trading network. Five homes that generated their own electricity from rooftop mounted solar panels on one side of a New York street were connected to five homes on the other side of the street, enabling them to sell electricity that was surplus to domestic consumption. The experiment demonstrated how domestic energy producers (prosumers) can trade electricity with their neighbours without the need for third party utility companies, catalysing further projects in the Netherlands, Australia and Tanzania. Blockchain technologies promise to remove the "middle man" in electricity distribution but depend on the mediation of software engineers.

The Transactive Grid (Back)

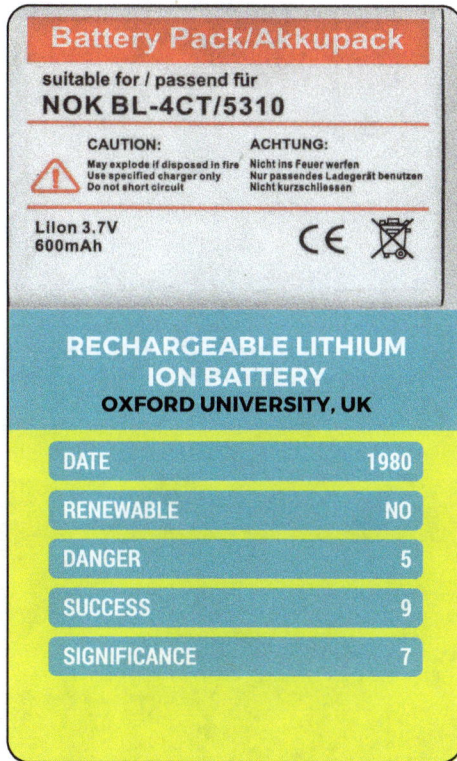

Battery Pack/Akkupack

suitable for / passend für
NOK BL-4CT/5310

CAUTION:
May explode if disposed in fire
Use specified charger only
Do not short circuit

ACHTUNG:
Nicht ins Feuer werfen
Nur passendes Ladegerät benutzen
Nicht kurzschliessen

LiIon 3.7V
600mAh

CE

RECHARGEABLE LITHIUM ION BATTERY
OXFORD UNIVERSITY, UK

DATE	1980
RENEWABLE	NO
DANGER	5
SUCCESS	9
SIGNIFICANCE	7

Rechargeable Lithium-Ion Battery (Front)

In 1980 John Goodenough demonstrated a rechargeable lithium ion battery cell. Lithium ions move from the negative electrode (anode) to the positive electrode (cathode) during discharge. First and popularly used in portable electronics lithium ion batteries are now a vital part of decentralised, solar energy systems. At the beginning of the 21st century companies like Telsa have pioneered the expansion of lithium ion batteries for domestic and industrial use. Yet lithium is a finite resource. So too is cobalt, the material used to make most positive electrodes in lithium ion batteries. As our dependency on lithium ion batteries increases new questions emerge about the terms and conditions under which these raw materials are extracted from the earth.

Rechargeable Lithium-Ion Battery (Back)

KIVU WATT METHANE PLANT
LAKE KIVU, RWANDA

DATE	2016
RENEWABLE	YES
DANGER	9
SUCCESS	8
SIGNIFICANCE	3

Kivu Watt Methane Plant (Front)

Harvesting the methane resource submerged beneath the waters of Lake Kivu., Kivu Watt is a unique experiment in industrial scale power generation. The project has proved an enormous success for Rwanda, fuelling the country's ambitions for accelerated economic growth. However rising methane levels are a result of the Lake Kivu's geology and pose a risk of explosion. The nature of the resource and the specificities of the location mean that this exemplar of clean, power generation is difficult to replicate.

Kivu Watt Methane Plant (Back)

THE SOLAR SUITCASE
ABUJA, NIGERIA

DATE	2008
RENEWABLE	YES
DANGER	0
SUCCESS	8
SIGNIFICANCE	8

The Solar Suitcase (Front)

The Solar Suitcase is a portable power unit that 'aims to improve health outcomes for childbearing mothers and their families by supporting health workers with equipment powered by the sun.' It was designed in 2008 by Laura Stachel, an American obstetrician, and her husband Hal Aronson, a solar energy educator, to provide obstetric care during frequent blackouts at a hospital in Abuja, Nigeria. In 2009 it won $1000 in a University of California Berkley innovation contest and a nonprofit company 'We Care Solar' was founded to promote it. The Solar Suitcase is assembled in California and has been deployed as part of the emergency response to natural disasters in the Philippines, Nepal, and Puerto Rico, and the Ebola outbreak in Sierra Leone and Liberia.

The Solar Suitcase (Back)

DATE	
RENEWABLE	
DANGER	
SUCCESS	
SIGNIFICANCE	

DIY Card Template (Front)

ELECTRO TRUMPS

DIY Card Template (Back)

NOTES

1 See Nye's 1998 anti-biography of Edison, which argued that there is no unitary Edison to explain or account for.

2 Top Trumps was launched and marketed in the UK by the Dubreq games company. The game's success over the next decade saw Dubreq acquired by the multinational company Waddingtons. In the 1990s the Top Trumps brand was bought by Winning Moves, a UK-based global games company.

REFERENCES

Cross, J., 'The Solar Good: Energy Ethics in Poor Markets', *Journal of the Royal Anthropological Institute*, 25.S1 (2019): 47–66.

Dumit, J., 'Game Design as STS Research', *Engaging Science, Technology, and Society*, 3 (2017): 603–612.

Fischer, M. M., 'Four Genealogies for a Recombinant Anthropology of Science and Technology', *Cultural Anthropology*, 22.4 (2007): 539–615.

Fortun, K., 'From Latour to Late Industrialism', *HAU: Journal of Ethnographic Theory*, 4.1 (2014): 309–329.

Jasanoff, S., 'Technologies of Humility: Citizen Participation in Governing Science', *Minerva*, 41.3 (2003): 223–244.

Kelly, A. H. 'The Experimental Hut: Hosting Vectors', *Journal of the Royal Anthropological Institute*, 18.1 (2012): 145–160.

Latour, B., and S. Woolgar, *Laboratory Life: The Construction of Scientific Facts* (Chichester: Princeton University Press, 1979).

— *The Pasteurization of France* (Cambridge, MA: Harvard University Press, 1993).

Lomas, R., *The Man Who Invented the 20th Century: Nikola Tesla, Forgotten Genius of Electricity* (London: Headline, 1999).

Marres, N., 'The Making of Climate Publics: Eco-homes as Material Devices of Publicity', *Distinktion: Scandinavian Journal of Social Theory*, 9 (2008): 27–45.

Marres, N., 'Testing Powers of Engagement: Green Living Experiments, the Ontological Turn and the Undoability of Involvement', *European Journal of Social Theory*, 12 (2009): 117–133.

Nye, D., *The Invented Self* (Odense: Odense University Press, 1998).

Perlin, J., *From Space to Earth: The Story of Solar Electricity* (London: Earthscan, 1999).

Pinch, T., '"Testing – One, two, three… testing!" Toward a Sociology of Testing', *Science, Technology and Human Values*, 18.1 (1993): 25–41.

Reno, J., 'Managing the Experience of Evidence: England's Experimental Waste Technologies and their Immodest Witnesses', *Science, Technology, & Human Values*, 36.6 (2011): 842–863.

Shapin, S., 'The House of Experiment in Seventeenth-Century England', *Isis*, 79.3 (1988): 373–404.

Torres Assis, A. K., *The Experimental and Historical Foundations of Electricity* (Montreal: C. Roy Keys Inc., 2010).

Tsing, A., and E. Pollman, 'Global Futures. The Game', in S. Harding and D. Rosenberg, eds., *Histories of the Future* (Durham, NC: Duke University Press, 2005), pp.107–122.

7

INTERVIEW: THE ANTHROPOLOGY OF ENERGY

Dominic Boyer interviewed by James Maguire

PREFACE

WHILE FINISHING HIS PHD, JAMES SPENT SEVERAL MONTHS AS A VISITING FELLOW at the Department of Anthropology, Rice University, Houston, Texas, where Dominic played host to his visit. Both share an interest in energy as an object of anthropological attention, as well as in the analytical forms being generated within studies of energy in anthropology-cum-STS collectives. Although James and Dominic are situated in very different institutional and national political settings, the politics of intervention in energy worlds has become an ongoing point of conversation between them. The piece that follows is, one should note, decidedly non-experimental in its form and sits in sharp contrast to the chapters that have preceded it; it might accordingly be understood as an experiment within an experiment, or a disruption to the experimental format, perhaps. Either way, it is also a shift that reminds us that experimental formats are not experimental for their own sake, but for the articulations and resonances they can generate. It is here that we feel a more classic 'interview' form can do a type of work that doesn't need to be experimented with.

INTERVIEW

JAMES: I'd like to start us off by asking about your career trajectory as an anthropologist, and how you came to study energy in the first place?

DOMINIC: For most of the early part of my career as an anthropologist, I was interested in questions of media and knowledge. Both of my first two major research projects focused on the ethnography of journalism. The first one traced what happened to East German journalists after German reunification. It was a post-socialist study, as well as a study of professionalism in transition. The second project was really about how digital information and communication technologies have reshaped the practice of news journalism over the past 30 years. Neither of those projects had anything to do with energy and environment issues, so you're right to ask what might have caused, or catalysed, that shift, and I think above all it was serendipity.

Cymene Howe (my partner) and I both got job offers from Rice University in Houston, Texas. We wanted to work on a project collaboratively. We also wanted it somehow to connect to life in Houston, as well as to be within a fieldwork range that was not too difficult to manage, since we had small children at the time.

Almost nobody knows anything about Houston, apart from energy, particularly oil and gas. So, we started looking into the anthropology of energy and saw that while there was already quite a lot of work brewing on oil and gas, and even nuclear energy, there wasn't much out there that focused on renewable energy. This was striking given the intensification of public discussions surrounding climate change and energy transition. We thought, 'Okay, maybe there's an opportunity here'.

We began scouring the world for places we could do research on renewable energy development and cast a really wide net. We thought seriously about projects in South America, also the Desertec initiative in Morocco. But what was happening in Oaxaca and the Isthmus of Tehuantepec immediately seemed very compelling. They were about to ramp up what would eventually become the densest concentration of onshore wind development anywhere in the world, and this was happening in a country that was increasingly struggling with a reputation as a vulnerable or weak state. Calderón [the president] was fighting the narcos and that created so much chaos and misery, but he also instituted these really ambitious clean energy targets that, especially at the time, positioned Mexico as a world leader in energy transition. That was what caught our interest. At the same time, Rice University was starting a cross-campus energy and environment initiative as a gambit to attract Brazilian oil money, and the two of us became part of the 'Cultures of Energy' faculty working group at Rice. Even though most of this group didn't really do research on energy, they nonetheless thought that Houston was the perfect setting in which to stage a more intense conversation about energy in the humanities. A similar idea was brewing up at the University of Alberta, and together that's how the 'energy humanities' concept got started.

JAMES: So, it was partially the specificity of Houston as a place that generated your field site. It's not that common, to my knowledge, that the places where people are located stimulate their field sites in that way; usually there is a lust for the somewhere else.

DOMINIC: I would imagine that Denmark must also be an inspirational place to think about energy, because so much is happening there right now in terms of designing national energy transitions. Not just at the level of 'let's throw a few solar panels here or there', but rather the Danish imaginary is 'let's try to design a zero carbon, or at least a low carbon national economy'. That's a pretty bold and fascinating project.

JAMES: Yes indeed. But it wasn't the specificity of Denmark – where I'm now living – that drove my interest in, and subsequent PhD studies on, renewable energy, but Iceland, where I initially did my Master's on fishing cultures. Interestingly, the relationship between energy and data infrastructures first cropped up while I was in Iceland, and has since looped back into my post-doctoral work on the emergence of the data centre industry in Denmark. The ways in which renewable energy has drawn big-tech data centres (Apple, Facebook and Google) to the windswept landscapes of Denmark is an intriguing story.

DOMINIC: Right. And, of course, that contradicts the point I made earlier about the digital and the energetic not having a lot to do with another. *Au contraire*, they are increasingly interrelated with one another. Bitcoin, I think, is a good example of this; how the big data, or the data analytics era, has a vast consumption of energy. Maybe that's a theme to talk about more broadly, why energy is such an interesting, indeed optimal, way to think about current and future forms of modernity?

And we weren't the first to arrive at this realisation by any means. In the course of reviewing the anthropology of energy literature back in 2008–2009, we discovered that the study of energy had an interesting periodicity to it. It was a topic the field turned to at moments of massive energy transition in society.

The first period was the 1940s work of Leslie White, and his effort to rethink the evolution of human civilisation through energy use. Although White was a kind of maverick, iconoclastic figure in a lot of ways, his work is still very interesting to read today. Although, not necessarily my cup of tea theoretically, his ideas about how new energy magnitudes unlock new cultural forms, and his weighing of nuclear versus solar futures, are actually quite contemporary in some ways.

Then, of course, in the 1970s we had the work of scholars like Laura Nader, but also many others who were ruminating upon the oil and gas crisis and the formation of OPEC at that time. The 1970s were a particular moment of energo-political rupture, during which the idea of pursuing an alternative renewable energy future flourished for just a brief moment in the United States, as elsewhere.

And now in our contemporary period, anthropological interest in energy definitely seems to be driven by the ambient crisis of climate change and the contradictions we are increasingly recognising between the massive, heavy energy, fossil fuel dependent lifestyle that we live today, and the idea of creating some kind of sustainable future.

JAMES: At the start of my PhD, I remember being at a conference at UCL in London in late 2013, talking to a senior scholar about my fieldwork in Iceland. I was a little bit taken aback when he asked me, in what I suppose you could call a less than curious tone, why I was even studying energy in the first place. 'What has this got to do with anthropology?' he asked.

I've got two questions. Firstly, have you ever experienced this? If so, does it still occur today? Then secondly, has it been difficult – as one of a small cohort of people – establishing energy as a legitimate field of research within anthropology?

DOMINIC: I can't say it's been difficult. The luxury I had of approaching energy in my third major project, instead of my first, is that I had enough 'bona fides' already in place to circumvent basic legitimacy questions like, 'What are you doing? What's wrong with you, thinking about energy?' I'm not saying it's fair, but people are often more willing to give your ideas audience once you reach a certain career stage.

The other thing that has been helpful, I suppose, to the legitimation (or more precisely, 're-legitimation') of this anthropological conversation about energy is that the past ten years have seen such a surge in the politics of energy on the ground, the grassroots and direct-action politics of energy. The events at Standing Rock became international news and attracted remarkable intersectional support. But there have been innumerable pipeline and anti-fracking protests across the world in the past decade, as well as a lot of activism focused around climate justice and environmental justice. Against the backdrop of public concern over greenhouse gas emissions and climate change, energy has become something that has seeped into the groundwater of public awareness and discourse in a way that makes the scholarly engagement of energy more salient too.

I never get the 'why energy?' reaction anymore. It's become obvious that energy is one of the most important issues of the twenty-first century. It wasn't necessarily that way in 2008 or 2009, when Cymene and I first began doing the background and ethnographic research for our project.

The anthropology of energy is, to my mind, fundamentally a collaborative venture, just like our project on wind energy has been, and just as the Cultures of Energy/CENHS project at Rice has been. At CENHS we're grateful for the great number of visiting co-conspirators we've had, yourself included – scholars, artists and activists – who've come and inspired us, and we're proud that a number of our fellows have gone on to start new energy humanities projects in their own right.

The collaborative or collective ethos in some ways also speaks to how energy studies or energy humanities is a subset of environmental humanities, and also draws upon that community's senses of relationality, response-ability and common cause, with their deep influence from critical feminist studies, eco-feminism, and so on.

JAMES: I was wondering what you think anthropology brings to the study of energy, and conversely, what we, as anthropologists and ethnographers, are learning from our study of energy?

DOMINIC: Maybe the obvious response would be to say that anthropology, as an intellectual space in the human sciences, juxtaposes a commitment to ethnographic methods, an interest in social theory and a commitment to fieldwork. In that nexus, there is something special about what anthropology can offer, especially when you're asking questions about the local meanings of energy development, and about local conditions and contingencies, considering what the lived effects of energy infrastructure and use are, and what aspirations people might have living upstream and downstream of energy systems.

In terms of our own fieldwork in Mexico, the primary research question concerned who was going to control the process of wind power development, and whose interests would be materialised. We went to Mexico as great proponents of renewable energy development, and we left as great proponents of renewable energy development; but what we learned along the way was that it really matters how one does that development. It's entirely possible to institutionalise renewable energy in a way that's very extractivist, and that's perceived to be extractivist by not actually satisfying local expectations or by not offering much community involvement.

One of the perspectives that anthropology offers in such a situation is an interest in discovering what it would mean to involve communities in processes of energy development from the beginning, and what it would mean to make sure that the kind of benefits that flowed from these projects actually achieved local social aspirations, which are of course different from place to place around the world. In a time in which the world really needs rapid development of renewable energy resources, that seems to be a very positive benefit of bringing anthropology into the mix.

On the other hand, I think the question about what energy brings to anthropology is a really fascinating one. Energy belongs to an ensemble of interests and attentions that have been motivating anthropologists of late. Including, of course, infrastructure, materials and multi-species relations. In other words, 'more than human' approaches in anthropology are becoming more central to its major conversations.

Anthropology is in an exciting moment in terms of decentring its own anthropocentrism, without losing, I think, the storylines that confirm humans have exceptional capacities and influence as a species. We've seen this in the Anthropocene conversation, the ways in which we've changed the conditions of possibility for the life of every other species on the planet; that's not something we should forget about.

But energy has given us a way to grasp modernity in different terms than, say, the growth of industrial economies or the spread of a certain set of philosophical ideas or values. My work is really focused on how to rethink political power, and power in general, through the lens of energy. What inspires me theoretically in this space is the work of Timothy Mitchell, and, of course, Hermann Scheer. Both helped to reveal how political institutions, commonly considered to be composed of ideas and norms, are also enabled through certain arrangements and infrastructures of energy.

JAMES: Yes, it seems to me that there's a clear connection between your work and Mitchell's way of thinking about the relationship between materiality and politics. Both address, in some sense, the relation between modes of fuelling society and modes of governing it. I think what's also nice about Mitchell's work is that it brings the dialogue between anthropology and STS into much closer alignment. Especially as energy becomes a lens through which to think about the multiple connections between infrastructures, the nonhuman, and politics.

Talking about analytical registers – some of which you just named a few moments ago – what do you consider the most exciting or most promising developments in the work that you're reading today? And, as an addendum to that: we talk a lot about concept development within anthropology and how concepts can have the power to help us to think differently about, and act differently in, the world. Do you think our conceptual contribution matters?

DOMINIC: Let me give you a little autobiography of my own analytical attentions. My work prior to energy on knowledge and media had me closely attuned to phenomenology, and I think phenomenology probably has been my most constant muse throughout my career, even as I've oscillated a bit in my flirtations with the praxiological, psychoanalytic and semiological wings of human-scientific theory. When I got into work on energy, I didn't give up that central phenomenological impulse. But it opened up a new series of questions such as how best to conceptually map the juncture between energy and experience.

I would definitely not have come to thinking about what I've termed 'energopower' and 'energopolitics' without Mitchell's influence. His carbon democracy project remains, to my mind, something that everyone should read. Talking with Tim a bit when he visited Rice, I came to understand that he was resistant to the generalisation of the case he was reconstructing. As much as he is a political theorist, he's also a historian, at his core. He was deeply, deeply committed to the empirical specificity of that case, which I very much respect.

When I discovered Scheer, I saw the potential for a broader analytics of energy in his comparison of long – and short – supply chains of energy and the different kinds of economies and politics they empower. That's when I began to think about energopower as a category; not as a type of power exactly, but as a conceptual lens through which we might reconstruct certain genealogies of power in the spirit of Foucault's work on biopower. To be clear, I'm not trying to make an ontological argument; I remain a phe-nomenological anthropologist at heart, who thinks that our categories of understanding are necessarily partial, historical and relational.

In that respect, I've also been inspired by the appearance of multi-species scholarship in anthropology and by the resurgence of feminist theoretical approaches in the human sciences more generally. Those approaches are often seeking not to tell the big monothe-istic narratives about origins – how everything came to be – but are more often trying to immerse themselves, engage and be present in a contemporary set of circumstances and relations. To stay, as Donna Haraway puts it so wonderfully, with the troubles of our times. Most recently, feminist scholarship has been enormously inspirational to me.

JAMES: A little earlier you mentioned a word, or at least you drew out the spectre of a word: ontology. I was wondering about your thoughts on the ontological turn? Above, you talk of feminist and multi-species scholarship, both of which have their own particular relationship to this turn, or opening, as others have called it. But given that you don't use the word 'ontological' a lot in your work, I am wondering if it is a register that you're comfortable with anthropology deploying. In a nutshell, my question is: how sensitive are you to the vocabularies and practices that have emerged alongside this term?

DOMINIC: I've written elsewhere that anthropology really shouldn't be in the ontology business. Ontology just doesn't play to our epistemic strengths as a relational field science in my opinion. What I can say about the ontological turn is that the parts of it that I find the most helpful are the ones that are essentially taking alternative and especially indigenous world views seriously. Not immediately casting out the non-Western as being fabulistic, imperfect or hallucinatory. There's such a long history within all the different strands of Western/Northern theory of indigenous understandings being forced to play the role of Unreason. So, if the ontological turn means not casting out indigenous ontologies as being stories people tell to make sense of a world they don't fully understand, then I am completely on board with that.

I like the move to take seriously the possibility that the Global North has not figured it all out. There are different cuts into the apprehension of the universe and we can learn from all of them. This is just to say that I'm someone who loves phenomenology. And conversely, I'm just sceptical of all ontologies, Indigenous and Western, to tell you the truth, in the sense that they are always narratives designed to overlay sense on the incomprehensible. Thus, I don't find the current continental philosophical takes on ontology all that illuminating. They can be very clever but I increasingly find myself thinking they are fairly fragile knowledge forms for use in the world, and are usually speaking some language of dominion.

So, while I'm not *against* ontology, you are right, it's not a practice that particularly appeals to me. It isn't very self-aware about its own conditions of possibility and propensity toward fantasy constructs. The reason that psychoanalysis, for example, is enduringly generative for me is that it's the one area of social theory that takes desire seriously. It's also the only theory that takes hallucinations and irrationality seriously. I look around the world and I see so much hallucination and irrationality regarding energy right now that for me, it offers a much more 'rational' point of analytic departure than ontology.

JAMES: At the same time there seems to be a series of associated terms that comes with the vocabulary of the ontological turn; for example, practice, enactment and multiplicity. It strikes me that scholars, particularly anthropologists, seem to be growing more comfortable in their adoption of such analytics, and I've been speculating as to why this is the case? Perhaps they work as kinship terms, given that many consider ontology to be a term too historically loaded to traffic in? But there seems to be some purchase in thinking ontologically, particularly when it comes to energy. As we know, energy evades clear categorisation, and so it becomes problematic to think of it solely in terms of the material, the conceptual, or even as conversion or measurement. It can be all of these things, although contingent upon the practices that give rise to it in any one setting. We also know that energetic materials are always undergoing change. So, in my own work on geothermal production, for example, energy is multiphasic, state shifting under varying conditions of extraction. Thinking of ontology as a method was a way to try and grasp energy materially – through its practices and infrastructures – as well as to think of material energetically, as subterranean fluids phase shift into different forms. It's in this sense that I think an ontological approach can be helpful, as a way to think performatively about how energy disrupts and overflows our categories and methods. At the same time, this cultivates a sensitivity towards the relationship between energy forms and their politics and ethics. I really like your reference to hallucination and irrationality – of which there is plenty going around – partly because it's an inversion of more traditional tropes about which types of peoples possess such characteristics. Desire is super-important in all of this, but I'm wondering what an ethnography of energy desire would look like? The ontological opening, as Marisol de la Cadena talks about it, leaves space, I think, for thinking about methodological approaches to such questions.

Along with Imre Szeman, you've been very much involved in what's been called energy humanities. This seems to have developed into an interesting cross-disciplinary space for thinking and practising energy transitions. I was wondering if you could say a few words about the work that energy humanities does as an academic space, and as a space of intervention?

DOMINIC: I agree, and I'm glad you brought up that aspect of it, because I do think that this is an area of scholarship in which, to my mind, the philosophical and the activist poles have to be deeply enmeshed in one another. It's an area that calls for much practical philosophical work. In other words, putting these ideas to work in the world, and allowing the work we do in the world to re-inform our analytical practices.

Energy humanities is not a group of people who want to leave environmental humanities behind; rather, environmental humanities is definitely the womb of energy humanities. It really is the source. If anything, I think the idea of energy humanities is to create a space of attention and conversation within environmental humanities, to really focus on the contemporary paradox and impasse of an energy intensive modernity that is creating these sprawling environmental impacts.

In other words, while there are a whole series of interesting questions being pursued in the wider world of environmental humanities right now – toxicity, chemospheres, atmospherics, multispecies relations, environmental infrastructures – in some way they cannot shake the *problem* of energy, because energy, especially fossil and nuclear, is the wellspring of so many forms of modern activity and luxury. And even many proposed solutions to contemporary environmental problems are themselves energy intensive, which creates a kind of feedback loop. Green capitalism is a massive feedback loop, for example, reminding us that electricity and fuel are the constant beating heart of the modern world.

In a way, the message of energy humanities to the rest of the humanities is just to say that energy is such a fundamental enabler of almost everything else we are interested in engaging critically today, and it deserves to be forefronted, because without the critical engagement of energy we won't truly address the repetition of legacies and toxicities associated with the energy intensiveness of fossil and nuclear modernity. Whether that's the toxic sovereignty of settler colonialism, or the toxic chemicals leaking into so many aquifers, we're not going to be able to address these issues unless these energy questions are brought into our awareness, because it's the unspoken reliance on particular energy forms and magnitudes which is often enabling these processes.

And since you mention Imre and the Petrocultures group at the University of Alberta, it's no surprise that energy humanities collectives have sprouted up in places like Houston and Edmonton, and other places that are nexuses of fossil fuel energy. When Jón Gnarr – comedian, author, and former mayor of Reykjavík – was with us at CENHS, we joked a lot about Texas as a Mordor-like fossil fuel space. And if we learned nothing else from *The Lord of the Rings* it's that real reform has to begin in Mordor if you actually want widespread change to happen. That's the role I see for energy humanists in places like Houston and Edmonton: hobbits inching up Mount Doom.

On the other hand, what you learn in these spaces is that their attunement to energy means that they contain a lot of people who are already thinking about radical energy futures, too, which is really exciting. Houston has some very interesting entrepreneurial activity in areas like high voltage direct current infrastructure which, while not well known, could really enable renewable energy to take over from fossil fuels. So, I wish the Danes every bit of good fortune in terms of getting to the zero-carbon national economy first, I think that would be wonderful. I think energy humanities has its own work to do there in helping to map and enact those futures. But even in places like Houston that represent the twentieth century more than the twenty-first, there are opportunities to inhibit repetition by getting those invested in the energy status quo – including universities, by the way – more attuned to its consequences and its coming transformation.

Universities, in my experience, although you would expect them to be places that would be highly attuned to issues like climate change and carbon footprints, are often very un-attuned to them. So, there's work to be done on campus, too, in terms of drawing attention to these issues, as well as in terms of trying to find ways through the arts and other media to connect with broader publics.

JAMES: There's been an ongoing discussion for many years now about the role of scholars in the worlds they research. Conducting research in what some have called catastrophic ecological times, or times of coming catastrophe, has only amplified this discussion. As academics become increasingly committed to a world-making mode of research – a central concern of this *Energy Worlds* book – terms such as intervention and experiment are becoming more commonplace.

On the one hand, these terms retain a sense of hope and promise for the politics that they could engender. On the other hand, they can be considered a little troubling, given the history of, say, experimentation. Could you say a few words about energy interventions and experiments as a question of method and/or politics?

DOMINIC: If I were to flag an area in which I'd love to see energy humanities get much more aggressive and creative, it would be in terms of actually engaging with energy experiments. I'm not a techno-utopian by any means, so there are limits to what we can expect from technology. Experimental practices also include engaging with the politics of energy and land. We've done some good work here at Rice around pipeline protests in Texas and in the Dakotas, for example. We're trying to increase awareness of the pipeline network as a massive infrastructure of power. Strangely, even when these pipelines rupture, they get very little media attention, and their invisibility is reinforced.

Those types of energy engagements, I think, are really important, as well as working with people who are on the frontlines of creating low carbon homes and low carbon

cities. We've got some great folk in architecture and urban design here who have ter-
rific ideas about how to redesign Houston so that it'll be less vulnerable to the massive
storms that will continue to come as long as our climate is moving in the direction that
it is. While there's plenty to be depressed about, I am still hopeful that in the long run
some good will come of this work.

I don't know whether this is something you've mused on, too, James, but I believe
we do have the technology right now to put an end to petroculture if the political and
cultural will were there. Even though our solar, wind, and battery technologies may not
be all that we would hope, they are already enough to create a non-fossil-fuelled economy
and energy infrastructure.

It would take a lot of work, it would take a lot of money, but it can be done. The
question is whether a non–petrocultured world would be similar to what we have today
in terms of practices and magnitudes of energy use. This relates to the kind of extractivist impulses that we've already talked about. Do we want to continue living in a world
dominated by extractivist logics?

This question really interests me; is it just about energy transition or is it about some
more profound kind of social and political transition. This is why I love Hermann Scheer;
he was a theorist of political transition as well as energy transition and was himself an
activist and a politician, who helped write the feed-in tariff law that made Germany into
a solar energy powerhouse, despite having so little sunlight.

JAMES: Just for clarification, Hermann Scheer is a figure that you have come to be interested
in through your research in Germany, is that right?

DOMINIC: Certainly the fact that I have been attuned to Germany for many years had
something to do with it. But I really discovered his work when I turned to thinking
about energy and power. He was someone who was thinking through the interdepend-
ency of contemporary energy infrastructures and state formation. In other words, the
'statist quo' – that's kind of a pun, but I like it – the statist status quo is something that is
linked almost inevitably to the long supply chains of nuclear and fossil fuels, according
to Scheer. Much of contemporary colonialism and globalisation is about keeping those
fuel sources moving smoothly, regardless of local interests and local livelihoods. That
analytical move, I think, is one we all should be practising.

I'm not saying that Scheer's solar utopianism is necessarily entirely reasonable, and
he has his own cornucopian fantasies built into his conceptualisation of solar power,

but, at the very least, I think we should all listen to the point he makes that this has to be about more than just energy transition. This has to be about radical social transition too. What we learned in our wind power research is that it's quite possible to maintain almost everything else from a colonial modernity minus the fossil fuels.

It was very disturbing to see so many communities coming out to fight against wind energy, not because they were against it in principle, but because they were so opposed to how the development was being done with so little regard to local histories, local interests and livelihoods.

JAMES: I think I've seen something very similar to that in Iceland, a place which you're also quite familiar with. Here we saw a scaling up from one renewable energy form (hot water for local heating) to another (the industrial production of steam to supply the electricity needs of the aluminium industry). That extractivist logic is still very present at the heart of renewable energy production. It's not just a question of the drilling technologies used in geothermal energy – legacy technologies from oil and gas – but it is also a question of the economic and political models embedded within such extraction methods, particularly the logics of fossil capitalism. This has had other types of effects in Iceland, where we are seeing the production of anthropogenic earthquakes as a consequence of such extraction methods. In many ways it is a cautionary tale: What do we need to change when we're thinking about transition? Of course, it's much more than the energy, as you say – it's also the forms of thinking and imaginaries that come with energy production, distribution and consumption. Maybe the question we should be asking is how do we de-fossilise our concepts?

DOMINIC: Beth Povinelli was fond of saying, while shrugging her shoulders: 'we're all late liberal subjects'; and she's right, but we're also all petro-subjects, too, and nuclear subjects. As we contemplate and act, and try to make different futures, I think we have to understand our inheritances and to be pretty ruthless in terms of interrogating where our ideas and affects concerning 'the good life' come from; which is mostly through the pleasure circuits and fulfilments of fossil-fuelled modernity.

JAMES: Finally, two questions about energy politics in the US. Firstly, could you say a little about our common energy futures, where might we be heading? Secondly, could you comment on the future of energy scholarship, as you see it?

DOMINIC: What I hear in the US these days is a lot of, for lack of a better word, solar Hegelianism, the world-spirit inhabiting PV panels and breaking through the spiritless formalism of fossil capital. We faced a kind of retro-cycle with Trump and his fossil fuel cronies, but it was so laughably gerontocratic, with the ancient coal magnates on TV begging for mercy; it's so hallucinatory what they had in mind (clean coal!) – it's so ridiculous.

If you work at a power utility or if you work in investment capital, you can't take it that seriously; the expert class is betting on the advance of renewables. But every year we lost by talking about something as ridiculous as clean coal was a year we lost to do something better. So, the delay tactics have their costs too. I think we have to be honest about that.

Of course, I would love to believe the solar Hegelians that there's an inevitable progress towards renewable energy, but I think we understand as well that institutions matter. One of the obvious effects of what is happening in the United States right now is that the US will slip behind other countries, as it has since the 1970s, in terms of the development of these technologies and infrastructures. In one sense, I think that's fine, America had a good run as an empire; the twenty-first century will not end as the era of American dominance the way it began. The more Americans vote for people like Donald Trump to be their president, the faster that decline is going to move.

Like you, I see optimism – as a self-identifying alt-globalist – in the fact that America's retreat from energy transition has meant that other countries now have had to step forward. Sino-European relations are clearly going to be critical for the global energy future. Africa and Asia will be critical to defining the future energy mix. In terms of the future of energy and energy scholarship, I have become increasingly interested, as you know, in electricity.

On the one hand, electricity is treated today as a kind of a salvational force. The idea that we are moving away from a fuel society and towards an electric society. It's true that much of our electricity continues to rely upon fuel, but what people project for the future – this is the Elon Musk scenario – is that we'll have Tesla solar shingles on our roofs, Tesla cars in our garages that'll double as batteries, and we will be able to, essentially, live off-grid in a stable, reliable, yet still modern way. That is even better than the grid because the grid fails sometimes, gets knocked out by hurricanes,[1] but nothing will blot out the sun, right? Nothing will stop us from getting our energy directly, and, as the technologies advance, our future will be increasingly hyper-local and hyper-electric.

I'm interested in that vision and I think that electricity deserves more critical scholarly attention. I think that questions of storage are going to be really interesting, too, because all of the hyper-electric models require some kind of storage solution, which are things that anthropologists haven't thought a whole lot about. What does it mean in terms of cyclicity and temporality, to be attuned to matters of energy storage?

JAMES: Yes, electricity is a really interesting focus. What I'm learning in my current research on data centres is that while there are numerous reasons why companies like Apple and Facebook want to locate their data in a place like Denmark, what is key is energy, and particularly big-tech's desire to green its data infrastructures. But there's also a series of ancillary energy projects hovering around the edges of the data centres. For example, Apple invested a huge sum of money into a local methanation project [converting carbon dioxide from the air into methane]. In a way it's like the holy grail of renewable energy battery storage. I like the idea you paint of the world becoming more hyper-electric and hyper-local. It feeds directly into some ideas about how to transform the data centre industry away from the tech giants and towards the possibility that local communities could use data centres as both data and energy providers if these experimental technologies become scalable.

DOMINIC: This brings up some fascinating issues about what is involved in load balancing. How do you orchestrate a national system of load balancing, or an international system, and still involve individuals, and families, and so forth, in those projects? How are those systems coordinated in real-time? I'm imagining that's going to be another very interesting area of research, too.

JAMES: Yes. But maybe these are conversations for another day.

DOMINIC: Indeed, but just to say that I'm really glad you're working on data centres, because I think we're just beginning to become aware of their energopolitical, if I may say, place in the world.

NOTES

I   At the time of preparing this book for print, a hurricane hit Houston. Extreme weather conditions and grid failure led to severe power outages and clean water shortages in parts of the city.

8

CONCLUSION

James Maguire, Laura Watts and Brit Ross Winthereik

THE CHAPTERS IN THIS BOOK BEGAN AS SINGLE AUTHORED COMPOSITIONS WRITTEN IN preparation for a workshop on energy research. During the two-day workshop in Copenhagen, the contributions took on a new life as participants rethought their work through a series of conceptual prompts about energy and experiment. Towards the end of the second day, we, as workshop hosts, invited participants to find fellow travellers within the collective with whom they could experiment and recompose their texts. It was challenging to find new constellations within which each author's work could reside, and eventually be rearranged. This volume is the result of that challenge: multi-authored texts within a multi-edited volume – an experiment in method and form.

Thinking and practising experimentation is nothing new within anthropology and science and technology studies, both of which have spent decades characterising and conceptualising experiments as they unfold in various settings. Whether it be laboratory setups (Shapin 1988; Latour 1987, 1988) or studies of experimental situations (Jensen 2010; Marres 2013; Winthereik, Johannsen, and Strand, 2008, Whitley 2000) or the appropriation of the term 'experiment' for creative methodological means (Lassiter and Campbell 2010; Marcus 2000, Rheinberger 1997) – helping to rethink social science methods, for example – what we want to hold onto is that acting experimentally can, and should, put the very definition of the term at stake.

Key here is that being experimental does not preclude critical engagement. Experimentation enlivens critique, shifting its mode from deconstructing to doing, rearranging and recomposing our stories, methods, and politics in the process. As others have argued, doing things otherwise is both an ontological and a political project (Haraway 2016; Verran 2014; Mol 1999; Povinelli 2016). The concept of experiment is the thread that binds all the chapters together. And it is the power of stories that holds this thread taut. Stories are powerful vehicles for communicating and making ideas. They are, as Donna Haraway has reminded us, more than just words: they are material-semiotic performances that help make worlds legible, desirable and accountable. Stories tell the world, but as this and other books show, they tell it differently depending on the material form of their transmission (Jungnickel 2020). Following the Introduction, Chapter Two 'The power of stories,' articulates the book's overall concern with the politics of storytelling. In the chapter, found ethnographic artefacts are the prompts that weave energy infrastructure stories together. The stories offered invite us to reflect on how each energy world is connected to other times and places through its particular narrative threads. The stories contain, assemble and stitch together the energy generated at each of these experimental sites. What the chapter emphasises is that while we might begin with the understanding that energy infrastructures are *in* places, we come to understand that energy *evokes* place, too, and that this evocation is not only a process, but also a highly political act of telling energy stories.

Chapters three and four, on propositions and theses, attempt to address what the call for doing things otherwise might actually entail, in the fields we study as well as in our own writing and collaboration practices. They are what one might call, in a more traditional sense, the 'political' chapters. But in the spirit of this book, each of the chapters finds a way to be experimental with the term experiment in an effort to assess what the political forms of the 'otherwise' might be.

In Chapter Three, 'Propositional politics', the 'otherwise' is an invitation – a proposition – to engage with various others that have traditionally been silenced in the articulation of who or what they are, and how they approach contentious problems. In analytically charting a move from experiment to proposition, this author collective suggests a less certain mode of engaging with and relating to human/non-human encounters. Rather than view the political practices of others as experiments that can succeed or fail in accordance with more traditional renderings of what count as political outputs, the authors argue that a propositional approach to politics has the power to open up the world in ways that are both ambiguous and promising. Instead of a search for ontological clarity and unambiguous responsibility, the authors advocate for an exploration of ties between practices and logics that might otherwise seem incommensurate. In this sense, a proposition is a future oriented, speculative provocation that takes various others seriously, whether they be seismic energy releases in Iceland, pacts around water in Chile, or meetings between Aboriginals and Western engineers in Australia.

In the case of Chapter Four, 'Five theses on energy polities', the otherwise finds its form in the genre of philosophical treaties, probing the ways in which spaces for political intervention can be carved out through brevity. As such, this chapter works against one of the central characteristics of deliberative politics: its slow, processual nature. The thesis is a political experiment whose form has arisen under duress, where prevailing power structures allow for little else other than short, incisive provocations, and proclamations against ruling bodies. In a way, the form of the document is closely connected to its environment. It is a protest form that resists a particular state of affairs through a provocative set of ideas. This chapter attempts to do something similar, by making connections between energy environments and the politics that they might afford. It does so by trying to grasp the not-quite-political. This 'not quite' doesn't mean *less* political, or doing politics outside its traditionally conceived bastions – as, say, technopolitics has done – but instead involves grasping the ephemeral nature of polities in becoming. These nascent and fleeting political forms can emerge through settings and constellations that might not otherwise be activated. The extent to which the ambiguous form of the thesis has any political agency, is an open question, of course.

Lest we forget, our present is one where school children and young people are our leading experimenters-provocateurs, resisting, as Isabelle Stengers puts it 'a future that presents itself as probable or plausible' (Stengers 2000). In multiple places round the world – both in physical and digital spaces – young people are speaking truth-to-power as they protest against the breaking of implicit intergenerational compacts, and forge new alliances with one another and with the nonhumans around us. They offer a form of inspiration that reinvigorates a project like this one, encouraging us not to tone down our experimental impulses, but to ramp them up in ways that entreat us to rethink our modes of connecting to their channelled fury. Although the experiments presented in this book run the risk of paling by comparison, we find more encouragement than not. And we also see an alliance of sorts. If part of what young protesters are resisting is the institutionalised attitudes and power structures that have produced a coming climate cataclysm, then what we offer here is the ambition to experiment and imagine together different ways of doing energy worlds within and beyond academic institutions.

It is here that Chapters Five and Six come into their own. Not purely as pedagogical tools – although they can also be used this way – but as imaginative fodder that inspires, we hope, alternate stories and modes of action. Chapter Five, 'Unda: A graphic novel of energy encounters' is an interruption to classical modes of writing and storytelling. It is also, particularly, about making energy more relatable through its embodiment. The protagonist's body is itself malleable, and altered by invisible electromagnetic radiation. Through her changes we are invited to see energy and electricity. Shapeshifting energy is literally contained in her changing form, creating a figure we can play with and whose presence invites alternate stories. This is a form that dazzles, where the reader is invited along *to feel and be affected by* the power of alternate energy worlds.

Chapter Six, 'An energy experiment: Tests, trials and ElectroTrumps' draws us into energy worlds through the form of a card game. Its experimental nature resides in both its knowledge form (as a card game), and in the way card players are invited to read and rework the materials. Energy and electricity infrastructures – often difficult to grasp – become as simple and as relatable as a basic card game. Energy gains a material shape; it becomes malleable to thought and analysis as players are invited to design their own cards that inspire new questions. At the same time, it is also a form of public engagement. What this translates into, for us, is the desire to develop novel opportunities for reimagining energy futures. As Lucy Suchman (2007) reminds us, reconfiguration, generating new energy worlds from existing materials, does not require elaborate means or electrical devices, but can result from reuse and juxtaposition.

All the stories in this book highlight how energy overflows boundaries – material and imaginative. They remind us that energy cannot be destroyed, only transformed. Experimenting with new forms of storytelling gives us hope that, in this era of climate change, we can build flourishing energy futures.

REFERENCES

Haraway, D., 'A Game of Cat's Cradle. Science Studies, Feminist Theory, Cultural Studies', *Configurations*, 1, (1994): 59–71.

— *Staying with the Trouble: Making Kin in the Chthulucene* (Durham: Duke University Press, 2016).

Harvey, P., C. B. Jensen, and A. Morita, *Infrastructures and Social Complexity: A Companion* (London: Routledge, 2017).

Jensen, C. B., *Ontologies for Developing Things: Making Health Care Futures Through Technology* (Rotterdam: Sense Publishers, 2010).

Jungnickel, K., *Transmissions: Critical Tactics for Making & Communicating Research* (Cambridge MA: MIT Press, 2020).

Lassiter, L., and E. Campbell, 'What Will We Have Ethnography Do?' *Qualitative Inquiry*, 16.9 (2010): 757–767.

Latour, B., *Science in Action* (Cambridge, MA: Harvard University Press, 1987).

— *The Pasteurization of France* (Cambridge, MA: Harvard University Press, 1988).

Marres, N., 'Why Political Ontology must be Experimentalized: On Eco-Show Homes as Devices of Participation', *Social Studies of Science*, 43.3 (2013): 417–443. doi:10.1177/0306312712475255

Mol, A., 'Ontological Politics. A Word and Some Questions', in J. Law and J. Hassard (eds), *Actor Network Theory and After* (Oxford: Blackwell, 1999), pp. 74–89.

Povinelli, E., *Geontologies: A Requiem to Late Liberalism* (Durham: Duke University Press, 2016).

Rheinberger, H.-J., *Toward a History of Epistemic Things: Synthesizing Proteins in the Test Tube* (Stanford, CA: Stanford University Press, 1997).

Shapin, S., 'Following Scientists Around', *Social Studies of Science*, 18.3 (1988): 533–550.

Stengers, I., *The Invention of Modern Science* (Minneapolis and London: University of Minnesota Press, 2000).

Suchman, L., *Human-Machine Reconfigurations: Plans and Situated Actions* (New York; Cambridge: Cambridge University Press, 2007).

Verran, H., 'Anthropology as Ontology is Comparison as Ontology',

Society for Cultural Anthropology, 13 January 2014, < https://culanth.org/fieldsights/anthropology-as-ontology-is-comparison-as-ontology>.

Winthereik, B. R., N. Johannsen, and D. L. Strand, 'Making Technology Public: Challenging the Notion of Script Through an e-Health Demonstration Video', *IT & People*, 21.2 (2008): 116–132.

www.ingramcontent.com/pod-product-compliance
Lightning Source LLC
Chambersburg PA
CBHW080648270326
41928CB00017B/3233